TIME TRAVEL

Fact, Fiction and Possibility

D0732777

TIME TRAVEL
Fact, Fiction and Possibility

Jenny Randles

BLANDFORD

A BLANDFORD BOOK

First published in the UK 1994 by Blandford
A Cassell imprint
Cassell plc, Villiers House, 41–47 Strand, London WC2N 5JE

Distributed in the United States by Sterling Publishing Co., Inc.
387 Park Avenue South, New York, NY 10016–8810

Distributed in Australia by Capricorn Link (Australia) Pty Ltd
2/13 Carrington Road, Castle Hill, NSW 2154

British Library Cataloguing-in-Publication Data
A catalogue entry for this title is available
from the British Library

ISBN 0-7137-2402-1 (hardback)
    0-7137-2404-8 (paperback)

Typeset by Method Limited, Epping, Essex

Printed and bound in Great Britain by
Biddles Ltd, Guildford and King's Lynn

Contents

Introduction

Why does the future become the past?
Why does the future never last?
And will tomorrow become today?
Only to fade away
Just like yesterday?

I BELIEVE that I have travelled in time. I have done so on several occasions. This is not a delusion. I make this seemingly absurd comment as a statement of absolute fact. But I mean it in three, rather different ways.

For a start, everyone travels in time each and every day. We move from this moment 'now' to the next 'now' (which we define as the future, until it becomes the present and then ultimately the past) at a constant rate of one second per second. This readily demonstrable rate of progress never changes. It is an immutable fact of the universe. Or so most of us think. However, that opinion has been cast aside by science. Time, as Einstein first told us, is like everything else. It depends on where you are and what you are doing. It can – indeed does – change. To a young child a six-week summer vacation is a very different sequence of 'nows' to that experienced as the same period by an adult. Subjective time seems to depend on where and when you are just as much as science has found real, physical, measurable time to do. Without this amazing concept – introduced by Einstein's theory of relativity – much of our understanding of the cosmos would fall apart. We will return to this complex form of time travel later, but it is by no means the only one that we must contemplate.

In May 1991 – like millions before me and no doubt millions since – I travelled in both time and space, thanks to American Airlines. I took the long, gruelling flight from west to east across the globe taking me from Hawaii to Manchester, with brief-stopovers to change planes at Los Angeles and Chicago. This was a 16-hour trek. Because it was summer, had I left just after dawn I could, by simplistic definitions of time, have reached England before sundown and never witnessed a single night fall, because there was around 18 hours daylight to pass through. But it was not quite so simple as that.

In fact, I left Honolulu and chased the sun across the sky, spending two nights in the air. Indeed, I left the tropics and this mid-Pacific Island late on Saturday evening and made it home to Manchester very early on Monday. I saw two sunrises during a journey lasting no more than two-thirds of one day. In effect, I time travelled into the future.

This kind of juggling trick with the calendar and the watch is becoming more and more familiar to long-haul travellers as the world becomes a smaller place. The jet-lag that it induces is the price we pay for the astonishing progress that transportation has achieved of late.

A century ago this was all science fiction. Jules Verne wrote the concept into his story *Around the World in Eighty Days* because then, of course, it was a new-fangled thing and rather less understood or recognized by his readers, indeed even forgotten often enough by voyagers. In a three-month journey circumnavigating the globe they might reasonably miss the vagaries of the clock which become all too obvious nowadays when jet aircraft make that same trip so fast you cannot avoid the pitfalls.

In the not-too-distant future the speeds achieved by Concorde or the space shuttle will seem just as feeble as the speed of the Victorian horse and cart does to us now. When that day arrives and we speed around the universe at an immense rate of knots we will have new, time-related problems to confront us, which modern-day Jules Vernes already tinker around with in their science-fiction stories.

These new dilemmas will be a bit worse than getting home with your body clock shot to pieces and feeling rather tired. You would have more than a headache to frustrate you after taking the round trip to the star Alpha Centauri on a super-duper high-speed space plane. Following the journey (probably of around nine years) you would get home to discover that your great-great-great-grandchild was waiting to meet you at the spaceport! The problem is that, on earth, time will have run at a far faster rate relative to your space plane chronometers. Several centuries will now have passed and you would have turned into Rip Van Winkle. Sadly, it's unlikely that you could even claim compensation under the rules of your travel insurance policy. That is a description of what would happen. Not what might happen. It is an unalterable fact of life, the universe and everything. It certainly beats what I would call jet-lag.

However, even as these things are both time travel, by any broad definition of that term, what you probably think of as time travel is a whole lot different. This is actually zipping into the future, seeing what goes on there and then returning to the present with a full knowledge of what you just witnessed. In the case of H. G. Wells and his voyager in *The Time Machine* it was something pretty useless, like knowledge of a war between mutants and wimpish servants a few millenia after our civilization is pushing up daisies. You might anticipate something using time travel for rather more valuable means, like discovering the winner of next year's Derby was a 20-1 outsider so that you can bet all those life savings on it. Of course, in practice, again, it is rarely so straightforward. But at least, the evidence suggests, some sort of literal time travel of this type has also happened far too often for it to be ignored. I can vouch for that.

In fact, I have time travelled this way various times and will have much more to say about it all later in the book. But for now I will describe the first occasion that I remember it happening to me, because it was what really grabbed my interest and a quarter of a century later resulted in my writing this book. If, of course, cause and effect, are not an illusion! I am by no means the first, nor will I be the last, to discover the magical enticement and heady feeling of casting off the shackles of time. It holds so much fascination for so many of us, because, I suspect, deep down we all know that it is a part of life. We all time travel regularly. But some of us are just better at remembering our time trips when we come back.

Here is how it all started for me.

It was 8 March 1968. I was still at school, studying for my science 'A' levels with wide-eyed astonishment at each new fact that I learnt. I had begun to explore those strange culs-de-sac where science feared to tread, the dark alleys abandoned by the safe, respectable professors who thought such no-go areas a very bad place to be. If you trod there then you risked getting mugged by your own self-delusions. But I was 16 and naïve. I actually thought that being a scientist meant that you had the urge to face mystery head on, ask all the questions that you could muster and – hopefully – come out of the eye-ball to eye-ball confrontation with sanity intact and a few new ideas to play around with.

Of course, in truth, science is more often about preserving reputations, building foundations brick by brick and persuading the control board to fund next year's lab time. To do this your work must be published in an obscure journal somewhere or other and absolutely not be of any interest to ordinary folk who may read the *National Enquirer* or *Daily Star*. Still, I had a long way to go before my physics tutors at university made me understand any of that and, for now, I was simply excited by the prospect of new experiments uncovered in a dusty old book picked up from the library. The fact that they were impossible and heretical passed me by. That book, entitled *An Experiment with Time* was by an aeronautical engineer called J.W. Dunne. Forty years before I read his work he had analysed his own dreams. Somewhat mystifyingly, he had then concluded they gave frequent previews of tomorrow. If you learnt how to scan and record them, he said, then you could prove this to yourself. From years of experiment he had found this out. Now the question was not if we could dispense with the tried and trusted definition of time but, as we could, what exactly did that teach us about how the universe functioned.

The process entranced me more than any toy at Christmas. It was like being able to turn on the TV set and watch what was going to happen in the future. I was mildly curious as to why nobody at school had ever told me about this fascinating hobby, but what the heck. Dunne gave straightforward instructions, even if much of his book lost me in a sea of mathematics about 'Time 1' and 'Time 2' which were irrelevant anyway. As a would-be scientist there was only one course of action open to me. Try it out and make notes of what resulted from my experiment. Dunne was just my kind of researcher.

So I did. Dunne had suggested that you keep a notebook by the bed and train yourself to remember dreams, seeking out those which foretold the future. According to him these were about one in two so you should not have long to wait to prove him right (or, of course – a possibility I was more than willing to entertain with no little relish – prove him absurdly wrong).

I was not sure if I dreamt much, or if I did then how often. But Dunne said that everyone dreamt lots, and every night. Those who said otherwise were just no good at remembering them. Modern science has long since more than verified what at the time was an unproven statement. It came from a man most termed less than qualified to talk about the mind – except on perhaps the most important ground that (like the rest of us) he actually had one.

Dunne promised that if I persisted long enough the truth would out. But by chance, as I see now but did not spot at the time, I was highly fortunate with the chosen date to start my first test run. Two days later I was due to participate in a very exciting experience and this, it is now clear to me, was a critical factor. New stimuli entering your mind in the near future seem to trigger visions ahead of them, like energy waves rippling two ways across a pond into which you drop a heavy stone. Also, memorable events full of emotions do much the same thing. I was heading into something destined to offer both such advantages at once so it is unsurprising that they were to provide the raw material for my first trip into the future.

Early on Saturday morning as I woke feebly into the daylight I wrote up my second batch of dream notes. I was still very inexperienced at this and have often regretted that I was less complete than I might have been in doing this task. Still, I scribbled into my little blue notebook bought from the corner shop for the equivalent of 2p and wrote about the typically meaningless and silly thing that I had just dreamt: 'Saw paper factory by bridge,' I began. I had witnessed a fire associated with it. That was all that I remembered. But it was the only vivid dream that night.

Later that Saturday I took part in the Bogle Stroll, a sponsored walk for charity organized by Manchester University each year. After being ferried to Lancaster in a fleet of buses more than a hundred of us set off to walk the 80 km (50 miles) back to the city through the relative calm of the early hours of Sunday morning. It was, of course, a memorable, thrilling, unprecedented event in my young life. The images of it are etched strongly even now, so it ought not to be much of a surprise that they jumped tracks a bit by around 24 hours, as, evidently, they did.

About half way into this marathon, at around 3.30 am, a little group of us were walking through Preston and approaching the River Ribble. Suddenly we saw orange flames licking the sky. 'Fire,' someone cried, and before I knew what was happening we were scurrying around a little side lane leading from the bridge over the river to the source of the combustion. A tall building loomed into the darkness. It seemed to be ablaze. Then we all burst into giggles when someone pointed out that it was just the furnaces burning with the night shift and reflecting off the low-cloud ceiling. We shuffled away, past a little sign that announced this was a paper production mill.

I was unsure what this news meant. Something was tickling in my mind that it was important. The very act of writing down the dream notes the morning before had meant that I half subconsciously remembered what otherwise I would have certainly forgotten, but I was unsure. The proof lay 40 km (25 miles) away, in my bedroom in Manchester. All thought of the Bogle Stroll paled against this sudden overthrow of the laws of nature.

At the next check point, in Chorley, I gave up. Nearly 50 km (30 miles) was a fair distance to walk, I justified to myself. Shelter – the charity for homeless people – would not do badly out of my efforts. But I had to get home and see those notes. I had to know, not simply believe that I knew, that science fiction had just turned into science fact, exactly as a man 40 years before had promised me it would.

I flicked through the notebook. There were the words. I read them two or three times but they remained the same. I had dreamt about a paper factory on fire beside a bridge and then under a day later had seen exactly that in real life – except, of course, that it was never on fire. My dream had not witnessed the reality. It had previewed my mistaken perception of reality.

Dunne's experiment had worked in exactly the ratio that he said it would. My second night's dreams had provided a remarkable proof of seeing the future. I juggled odds about dream images of bridges, paper-mills and fires around in my head trying to work out what the chances were that this could just be a coincidence. Three years later, when I dared mention it at university to a relativity physics lecturer, 1 got a pitiful look and was told that billions of dreams happen every night. All of them are nonsensical but by sheer statistics every now and then one or two will seem to match up with a coming real event. This is not a premonition, I was told. This is mere happenstance, utterly predictable by the deterministic laws of chance.

I almost believed him. Possibly, had this been an isolated, one-off incident in my life, uncomplicated by many sequels or the countless more hair-raising examples I have investigated from other people's encounters with the unknown (something that my pompous professor would never have wasted time on) then perhaps I would have gradually persuaded myself that I was a victim of my own desires. I wanted time travel to be real, so I made it real. This explanation works – for about ten seconds – until I conjure up the memory of that sheer, cold certainty of the synchronization of dream image with reality. No amount of explaining after that makes any difference.

I have had so many of these wretched 'coincidence dreams' since 1968 that my towering pile of statistical odds long since toppled over. Precognition is the only solution that makes sense to me. Somehow or other – just as Dunne had told me all along – I had travelled through time. The world was now forever – and as it probably always had been – a very different place.

This was the experience that drove me towards the writing of this book. I owe a debt of thanks to J.W. Dunne and want to extend and modernize his pioneering efforts into a comprehensive review of time travelling.

To do this I must look far and wide. Whilst most books that do this sort of thing (what few of them there are) tend to lose the reader in deep debates about philosophy or nuclear physics (both unfortunately significant to the discussion of what time is) or else concentrate solely upon dreams that see into the future, I will be going well beyond that remit.

I will look at how the idea of time travel has captivated attention through novels, TV series, movies, etc. This may seem superfluous but in a field where experiments like Dunne's are not brimming over, a lot of interesting work is done through fictional creations; especially when quite a bit of that fiction is penned by astrophysicists or nuclear scientists!

I will also look at every conceivable type of time-travel evidence – from moving back into the past, witnessing scenes from long ago, to attempts to research past lives through creative hypnosis. We will show how some people move sideways in time – an odd-seeming, yet important, concept best left until real cases can be presented by me to illustrate it. Plus, of course, there are any number of ways in which the future has apparently been perceived – and not just through dreams of trivia like non-burning factories.

There will be some very surprising discoveries too as we pass along the way. The possibility of making time-travel machines is being seriously explored by science today. Where will this lead us? Is there any evidence of time travellers having come from the future when (and seemingly it is when, not if) we perfect this technology? After all, if we do make a time machine – ever – (be that in 1999 or 19,999) we should see signs of its appearance in the world today (indeed throughout recorded history). Do we?

Our search for evidence will lead us into curious territory, but it is a search that must be made. I will also have plenty of ideas for experiments and research methods that we can all carry out to test the possibilities of time travel. We are beyond the point where we need to ask if this happens – and into the area where we can try to find out why it does.

There are clues out there. My own very first experience showed one of them. I clearly did not preview the reality of a future event. If I had then my dream would have been of a paper-mill night shift, with its furnace blazing and shining off clouds. What I saw was one of three things – perhaps, the complete coincidence that science would have us believe it was, although I doubt that somehow. Maybe then a perception of some indeterminate future when the factory was really on fire, or indeed a parallel reality where this fire took place. Or, thirdly, as I suspected right away from that March day, a detection of my own future mental state – thus including my mistaken, temporary belief that I was witnessing a real fire.

I did not tune into a TV picture from tomorrow night, but received a telegram from my own mind 24 hours ahead of itself, relaying a split instant of time when I first saw that fiery mass and put two and two together to make a miracle.

In conclusion I will endeavour to provide you with the most up-to-date and rounded look at the concept of time travel – in fact, fiction, reality and possibility. It may not be scientifically erudite. It will not lose you in 'n' dimensional equations that only a handful of genius physicists comprehend. But it will focus on real events, actual experience, serious research and try to offer genuine insights into what seems to be going on.

This is a book I have wanted to read for a long time – but, since nobody else had written it, I guess I had to write it for myself.

Most of what we have learnt throughout our lives about the nature of time has been imposed upon us by our own self-deception. Frankly, the evidence suggests that it is little more than stuff and nonsense. Now that we are asking questions about this mysterious entity we are beginning to grasp some incredible truths. When it comes to travelling through time, there are as yet no rules and no obvious barriers. In fact, nothing is impossible after all.

Note: The use of an asterisk (*) after a name in the text indicates that the name is a pseudonym.

1

The Romance of the Future

I HAVE A THEORY about the real reason the *Titanic* sank. It is widely assumed to be because this marvellous vessel, whilst making its April 1912 maiden voyage across the Atlantic, hit a rogue iceberg between Southampton and New York. Being woefully bereft of lifeboat cover, it went to the seabed with hundreds of its passengers still aboard. What this conventional view does not take into account is the weight of all the undetected stowaways it must have been carrying. They were not recorded in any log-book. They could not be added to calculations about depth of passage below the water line. And none of them hung about when the ship started to go down. They were time travellers, visiting this one unique moment in history as keen voyeurs, just like crowds of people gathering at the scene of a terrible roadway pile-up, paying homage to the thought that it is not their turn to meet fate head on.

Virtually every TV series ever made about time travel has the hero and/or heroine magically visiting dramatic moments in history like this. The scientists sucked through 'The Time Tunnel', the young boy and his amiable cohort in the less accomplished US series 'Voyagers', each and every one lands up on the decks of the *Titanic* at some point in the episodes. There must be enough of them to have their own reunion to celebrate the anniversary of their escape. Indeed, as a child, it always mystified me more than how such travellers got there in the first place just exactly why they always managed to land slap-bang in the run-up to an earth-shattering event, not a week last Tuesday when nothing much happened. Of course, moderately flippant as this suggestion is, like most references to time travel in fiction it does make a serious point. If time travel ever becomes a fact, will future historians trek back to classic moments such as these, perhaps taking school kids on tours to learn about key events from first-hand observation, rather than from books?

If so, then wouldn't we know about it by studying in minute detail the records of tragic events and seeking out common denominators? What if you studied photographs of the lead-up to say, Custer's last stand, the *Titanic* going down and the crowd watching Babe Ruth hit that big home run – and in these pictures scattered across a century of time you found the same tiny face, unchanged by the passage of the years, but staring out at the action and looking suitably wide-eyed?

I have effectively just written the introduction to a science-fiction story about time travel and in doing so proposed a real experiment that somebody could do. I doubt very much that anybody has ever taken this idea seriously enough to try it out, in which case all those who scoff at my time travellers' reunion on the decks of the *Titanic* really don't have much of a case to offer until somebody does prove there is no such evidence! Proof that time travel is a reality may be out there already in hundreds of historical photographs. Such is how major scientific breakthroughs occur.

Serious speculation about time travel is merely 100 years old. Until then nobody really thought about the possibilities because nobody really thought about time. It just was (or is – or will be) and that was that.

In a sense Mark Twain wrote one of the first time-travel stories with his much-filmed tale from 1889, *A Connecticut Yankee in King Arthur's Court*. But here the time travel is a gimmick to set a story in the past, even though the modern origins of the traveller are important to what happens. The transfer across the centuries is dispensed with rapidly – the hero being zapped by lightning and just going there. A frequently copied mode was set up by all this, posing 'did it really happen or was it all a dream?' which had its ultimate moment of glory (if you can call it that) in the TV series 'Dallas'. Needing to bring back a major character who had left for good but had been persuaded to return to plug falling ratings, the entire previous series in which he had not appeared (indeed his character had been indisputably killed off) was turned into a dream, with the new series starting off where the one before had ended! This is what fiction writers call artistic licence and many a Hollywood or TV mogul has owed a debt to Mark Twain.

However, in this book his enjoyable romp was really only important in creating the concept of a time-travel story. It was H.G. Wells with his book *The Time Machine* in 1895 who really created the genre. Wells did at least make a passing attempt to justify what was happening. His hero was an inventor who forsook the more common avant-garde technology such as airships and submarines that were the staple of the contemporary science-fiction author, Jules Verne. Indeed, Wells showed his own inept 'skill' at prognostication by rather foolishly announcing that he dismissed submarines as any sort of practical possibility. Anyone daft enough to ride in one would suffocate!

No, Wells decided to invent a machine which could speed time up so that, whilst to someone aboard only minutes were passing, outside the world shot by millions of times faster. After a brief period travelling like this you could slow down to a stop and find that the world outside your machine had passed through countless centuries whilst you endured mere minutes. Amazingly, of course, we now know that Wells was foreseeing an actual reality in extraordinary detail. He had, in a sense, predicted by 20 years Einstein's progress in relativity theory, which later still was proven to be fact by quantum physics experiments. Here any thing (e.g. a time machine) need only make itself move at a substantial fraction of the speed of light to do precisely what Wells's invention was meant to do.

However, there was a major difference between Wellsian imagery and scientific reality to come. With Wells's machine you could flip a lever and head back home, eventually returning to a few moments after you had left. You might have spent a year in the future but to the traveller's friends in the 'present' you would have seemed to have been away only a short time.

This is something Einstein's theory would not suggest as a consequence of extremely high-speed travel. However, it is something that is today more reasonable than ever before as a consequence of further discoveries about the amazing world of sub-atomic physics. We will look at these later.

Equally, it is also something that real people alive today believe to have happened to them – not in fantasy but in actuality – and there are plenty of eyewitnesses to the extraordinary stories to back that statement up. We will also return to these later.

For now just accept that, crazy as Wells's fiction looked to be 100 years ago, everything we have learnt about the universe since his day has gradually eroded its fantastic nature to the point where it is almost too mundane to be called proper science fiction!

When writers wanted to emulate Wells there was, of course, a long gap before science caught up with his fantasies about time and, even when it did so from 1920 onwards, longer still before relativity, and later quantum physics, became even remotely comprehensible to 99 per cent of would-be scribes.

In this period the usual preference was to focus on a philosophical idea about time that took a lot of attention in its day. This utilized the fact that subjective time differs from person to person. Even Shakespeare saw this and had often commented upon it.

In 1970 I spent eight hours a day in a toy factory making screw brackets for what was laughingly called a holiday job. I had to create 20,000 in a week by turning a handle and pressing a little sheet of metal, then throwing the creation into a bin and starting on the next one within a few seconds. It was the most astonishingly mind-numbing thing I have ever done. These days robots do it a lot better because they do not get bored.

Although I only stuck that job out for a couple of days it took longer in subjective time than the months I later spent as a finger-print detective doing what, to me, was truly fascinating work. The adage 'time flies when you are having fun' is no old wives' tale. It is a reality. From it came the concept that we only experience the steady passage from one 'now' to the next because it has become a consensus belief system that we have been indoctrinated into since birth. Philosopher Henri Bergson in the late 1890s showed that children must learn to see time passing.

Misleadingly named primitive cultures seem a whole lot less time bound than we are. They don't plan each minute by the Filofax or mobile alarm. In addition, whoever they are and wherever they come from (be it Aborigines from the Australian bush, medicine men in an African wandering tribe or the Kogi Indians in the Amazon with their sophisticated grasp of ecological dangers) they share something important. They have what are called shamans – wise people, rather like mediums (indeed I suspect mediums are merely western civilization's own shamans). They are evidently capable of casting aside this mundane view of reality and seeing an eternal, timeless dimension which the aboriginal culture well encapsulates as the 'dream time'.

Several science-fiction stories have adopted this idea to suggest that time travel may be possible without any technological wizardry. Thinking is literally the best way to travel.

I used it myself in one of my short stories, *John Lennon is Dead* (1983). I had discovered a time anomaly – a mysterious death at an early concert by the Beatles (before they were even called by this name). It was at a club near where I was then living on Merseyside. Nobody seemed to know who the boy might be who had been crushed by the crowd. There almost certainly was a mundane, tragic solution, but it set me thinking about those time travellers arriving, departing again into the future and so never being traced after visiting the site of historical events. I walked to the spot where the incident had occurred a quarter century before and conjured up my story, as much as an exercise for my own imagination as anything else.

In the plot a teenage boy becomes obsessed with John Lennon after his (real life) brutal murder in 1980 and desperately seeks to find a way to stop it from happening. The youth's desire and the act of shutting himself away in deep depression within his room cut him off from the time flow. He loses the grip that this succession of nows passing inexorably by seems to impose upon us artificially. As such he finds himself back in an age before Lennon and the Beatles come together and loses himself in the wonder of this fantastic situation. Apparently he has many years to become associated with Lennon and the other Beatles and warn John not to move to New York. That simple act, he thinks, is all that is required to change history.

Gradually he realizes there is a problem that was unforeseen. Well before 1980, in fact in the very near future, the boy himself is going to be born – since he was (or will be) a teenager when Lennon is shot. What would happen then? How could he be in two places at the same time? He begins to suspect that he might disappear, or die at the moment of his actual birth and starts to leave memos to his future self, partly to warn himself of what might happen and partly to explain the truth to those ahead. He buries them in the real grounds at 'Strawberry Fields' – so beloved by John Lennon and epitomized in what was (or rather would be when it later came to be written!) this boy's favourite Lennon song. He hopes that somehow this will work. Of course, he also realizes that he is running out of time to warn John Lennon and can do it only by pressing through the crowd at a very early Silver Beatles concert on Merseyside when . . . well, you can guess the rest.

Even this very simple (and far from well executed) story shows you the extraordinary complexities of time travel. Almost everything you do has unforeseen dangers and each step you take provides evidence of the paradox that time travel itself must always be. Many brilliant scientists in the 1990s, whilst admitting that time travel seems to be a reality from experimental results and predictions arising from their quantum mechanical equations, none the less feel that there must be something they are overlooking because such paradoxes prove that time travel is impossible.

In its hoariest form, this paradox is usually framed quite simply.

Imagine you travel back in time to a point before your parents met and thus conceived you. Something you do stops them meeting, or, worse still, causes their deaths. You change history. But then, of course, you can never be born, and yet, if you were never born then how could you have travelled back in time to kill your parents and prevent yourself being born? Are you born or not born? The question is insoluble and for many this simple 'thought experiment' proves that time travel can never occur.

Professor Stephen Hawking, widely considered one of the greatest scientific minds of the century and the man who has come closest to solving the problems even Einstein failed to crack about quantum reality is a case in point. So well revered is he that this seriously debilitated man with the mind of a genius has made British TV adverts and had a number one best-selling book about cosmology. In 1992 when he asked for a part in the hit TV series 'Star Trek – The Next Generation' they actually wrote a script just so that he could appear 400 years hence as a holographic computer simulation on the spaceship's 'Holodeck' to pit his wits alongside other brilliant scientists from the past with the super-intelligent android, Data!

Hawking has said of time travel that there must be a law within physics as yet undiscovered, which acts to prevent it whenever the danger would arise from quantum events. This must occur to prevent the paradox that would otherwise occur. Or else, of course, as in my story 'John Lennon is Dead', the paradox never arises because events conspire to ensure that it does not. Here a loop in time closes so that my would-be changer of history not only fails to warn Lennon, but himself dies in an accident that resolves a minor historical riddle at a point just before he is actually born to prevent the same physical being ever occupying two different spaces at the same moment.

Such concepts are a feature of quite a few stories a lot better than mine. But I use it here to illustrate the dilemmas posed by this intriguing, but mind-warping subject we are marching boldly into.

Another example of the mental-travel genre is Richard Matheson's 1975 novel *Bid Time Return* (well filmed in 1980 as *Somewhere in Time* with Christopher Reeve and Jane Seymour). Here Reeve falls in love with a woman from half a century before whom he sees in an old photograph and he literally wills himself into the past by dressing in old clothes and removing every vestige of the modern world from his room. He succeeds, but forgets a coin that is accidentally lodged in his coat. When he finds this the bonds of time reassert themselves and he is snatched back into the future.

The movie's ending is rather happier, which works on an emotional basis.

There was also the rather curious 1986 film *Peggy Sue Got Married* which is well thought of, particularly by women, as it is unashamedly romantic in tone. Kathleen Turner (struggling valiantly to look 18), collapses at her school reunion (hallucinating? time travelling? drunk?' – we never find out) and gets to relive the past 25 years of her life and seek to make her love affairs work better the second time around. Of course, time has a few tricks up its sleeve as usual and the plot loopholes are rather larger than normal. But this was never really meant to be a time-travel story – just a story that happened to need time travel to be told.

When it comes to trying to fathom out time the stories really come into their own when the author is a scientist. One of the best examples is astrophysicist, Professor Fred Hoyle's *October the First is Too Late* (1966), which proposes time running at different rates in different spatial locations and time travel being possible by moving laterally, not in time.

He fleshes this out with some quite complex theorizing which, he explains in the book, he intends us to treat seriously and not merely as a story feature. Essentially he proposes that all times exist simultaneously in the cosmos rather like a huge shelf full of pigeon holes into which letters are slotted such as you find behind the reception desk at a hotel. Just as you might go along and ask the clerk to look for anything in the 'A' pigeon hole and then later I might go and ask for any under 'R', followed by someone who requests the 'M' hole to be scanned, this works fine because an external force (the clerk's consciousness) is viewing the pigeon holes.

Hoyle envisages how we become aware only of what is in our hole. A 'now' is illuminated for us by this act. But the 'M' hole might have access to knowledge about all those in the range 'A' to 'L' and anyone accessing the 'P' hole, for instance, would be partly aware of 'M's future.

It is rather hard to grasp any of this, but it is mathematically sensible and probably as good a way of trying to understand time as any other. In short, what it really boils down to is that it is our mind which imposes our one-dimensional perception of time onto what we experience – or as just the pigeon hole we happen to be in at that moment. However, at a more fundamental level all pigeon holes exist at once and are theoretically accessible, if we can find a way to step back from our blinkered reality – and as a result see them all lined up behind us side by side.

Once you enter the realms where a traveller does what Wells proposed from the start, go into the future, then back into the past, virtually at will, you become surrounded by that dreaded paradox. Whether in fiction or reality you have to find a way around it.

There have been many attempts. The incredibly successful *Back to the Future* trilogy of movies in the 1980s had some interesting plot twists as the time-travelling sports car zipped between time zones. Changing the past was a cornerstone of the plot but the paradox really did not phase the writers. This was, after all, not serious science fiction but pure entertainment, so it could happily overlook the apparent absurdity of A knowing that B was in danger of being killed a century in the past after reading a newspaper or seeing a photograph, thus travelling back to prevent this disaster and watching the photograph and newsprint change by magic.

In more structured attempts to think around such dilemmas one of two things usually occurs to avoid the paradox. Either it proves impossible to change history (as in my story) or else if history does change and the world becomes completely different, then because of that alteration nobody is aware of having jumped across the time stream or that there ever was a time stream that was different from the one they now perceive. In other words, A stops B from dying in the past, so B goes on to have a family and in the newly created present all of these people, their descendants and achievements exist. However, nobody (including the time traveller) is aware that there was ever a present where this fact was not the one true reality.

Both of these outcomes, unlike the 'back to the future' scenario, can be justified to some degree by present scientific knowledge. They have inspired countless tales of 'alternative' worlds, where Hitler won World War 2, the south beat the north in the American Civil War, the atom bomb was never built, and the British Empire still rules the world courtesy of giant airships! Indeed in his 'Q' series of novels about experimental time travel, beginning with *Seeking the Mythical Future* (1977), Professor Hoyle's son Trevor Hoyle has maintained the family tradition by creating alternative realities resulting from time meddling.

I think the best attempt to examine this problem comes in Ursula Le Guin's masterpiece *The Lathe of Heaven*. Here the character dreams of a slightly different world from the one he knew, then awakes to find that the world has literally changed to match the slightly out-of-phase dream world and nobody else knows this fact. Everybody behaves as if it has always been that way. Each time a dream comes the world is in a slightly different parallel reality upon the dreamer's awakening. We must ask from this whether the dreamer is creating the world anew each day or whether, as quantum physics suggests as a real possibility, there are a myriad almost identical but subtly different 'parallel' worlds and the dreamer simply travels sideways through space and time into a new one of these from day to day?

The paradox rears its head quite often in more recent stories and films. The clever 1980 movie *The Final Countdown* builds upon several real cases we will look at later in the book and projects a nuclear-powered warship, via a strange meteorological phenomenon, into the Pacific Ocean of World War 2 shortly before the Japanese sneak attack on Pearl Harbor, Hawaii. The fantastically armed modern crew finds itself with the ability probably to end the war three years early, but what will happen to the future world from which they have come if the US Navy changes history in this fundamental manner? The movie steers around the paradox effectively with some clever sub-plot twists.

In both the 1960s and 1990s series of 'Star Trek' on TV the time-travel theme and the paradox occur more than once.

The crew of the original *Enterprise* are projected through a portal in time in the award-winning episode 'The City on the Edge of Forever'. They emerge in gangster-infested earth in the 1930s where they face the dilemma of whether to rescue or allow to die in a fated accident a kind-hearted woman played by Joan Collins. This was one of the earliest sophisticated attempts to handle the time-travel paradox on film or TV.

A quarter of a century later, in the 1991 episode 'A Matter of Time' a man arrives on the new Starship *Enterprise* to announce to the crew that he is an historian from the twenty-sixth century travelling back 200 years to find out what it was really like on their ship over a 24-hour period that, he implies, has some great significance. As far as the crew are concerned this is a routine mission and, obviously, they desperately want to know what this man from the future knows that they do not know and what decisions they should take at appropriately critical moments which may, or may not, change the future. The visitor, citing the paradox problem in justification, rather mischievously toys with them but gives nothing away about what will happen, if anything, and what they should do about it in any event.

The story is, in fact, more subtle than this, with a quite clever denouement, best not spoilt by repeating it. But it shows the basic difficulty. Would future travellers have to be incredibly discreet in order to ensure they gave not the slightest hint and so provoke the paradox effect? Say they were merely to grimace at the point when they knew a button was due to be pushed that would have fateful consequences? That might be sufficient to make the wary person pause and decide not to push it, thus changing centuries of history and, as a result, the entire universe. Time travelling would not be for the faint hearted.

There have, of course, been many novels developed that use the paradox to good effect. Poul Anderson's *Guardians of Time* began as a series of short stories in 1955 and envisages a 'Time Patrol' for which people are recruited from all periods of history. They train at an academy in the pre-human, post-dinosaur Oligocene period of geological history. This fantastic complex is left there for half a million years so that any number of graduates can take the course. Then it is carefully destroyed so that when the area becomes settled by humans millenia later no sign will be left of this building from the far future. But could they remove *all* traces?

We follow a recruit to the academy being taught the laws of time. The paradox is uppermost. But, he is told, just minor incidents, e.g. killing one sheep by accident, will not fundamentally affect the future because as time passes genetic factors even things out. Furthermore, if you try to alter time you have to 'work real hard at it', because time has a fabric that snaps back at you. Stop John Wilkes Booth shooting Abraham Lincoln and, it is explained, the chances are someone else will do it and Booth will get the blame. You cannot defeat the paradox without a fight.

The 'Time Patrol' exists to scan for time anomalies – things in recorded history that suggest that a rogue time traveller may have fouled things up somewhere (or rather somewhen). Their job is to go in and clean up the mess and make sure time takes its normal course.

Forty years on from Anderson's adventure yarn science fiction has begun to cotton on to the quantum mechanical complexities of physics which, in truth, are far more incredible than anything H.G. Wells might have dreamt up.

For example, Ian Watson is a very clever ideas-writer. He has penned several stories about time, in which he often perceives the mode of travel to involve an altered state of consciousness. Here the mind can surpass the boundaries of both space, and consequently as physics suggests, time. *The Embedding* (1973) is a case in point.

But in *The Very Slow Time Machine* (1982) he develops the interesting idea that a time machine would need energy to operate. If it has to go, say, 20 years into the future from its starting point then it would first have to crawl slowly back 20 years into the past, gather up momentum like a catapult so it can fire itself forward using this accumulated power. His story analyses how people in the present might experience a future time traveller moving backwards through their continuum gathering up the strength to zip forward into the more distant future.

Probably the most extraordinary adaptation of quantum physics to the novel is Gregory Benford's *Timescape* (1980). Benford is a physicist working in the field and the ideas that he portrays in this well-designed story are all too credible.

He envisages an experiment in 1990s Cambridge where a scientist decides to use tachyons, a theoretically real set of sub-atomic particles that some researchers do believe to exist and have sought in their experiments with tantalizing results. Tachyons are like the many other sub-atomic particles with exotic names (such as muons and leptrons) discovered in the last half century from nuclear reactions. The one big difference is that, if real, as some theory and results indicate, they are travelling backwards in time! Using tachyons as a means to send a pulsing message through space (and, of course, time) the scientist aims to warn the past about the polluted mess the future world has become.

Meanwhile, back in 1963 a scientist in California finds that his experiments are being interfered with by strange signals. He knows nothing of the still undiscovered tachyons, of course, but starts to investigate. He finds that they seem to be spelling a message. As you might expect, this is then erroneously assumed to be coming from an alien civilization in space.

Benford very cleverly shows the way that real scientists would struggle with these problems and what the consequences might be of such two-way time travel of information only – i.e. nobody ever physically travels in time. As information time travel, with or without tachyons, is actually far easier to envisage and a lot more likely to be perfected in the near future, this is a unique and perceptive story. It also opens up a number of other questions we will address later in the book.

Visually, the hit TV series 'Quantum Leap' (1988) used this most effectively over its several year run. It was a departure from earlier time-travel shows because it deliberately limited the years in which it argued that travel was possible. It also avoided ending up at the scene of major historical events every week, tending to involve itself with mundane life, as indeed all the real time-travel stories do prove to be the case.

In the series a brilliant quantum physicist from the late 1990s

has perfected an experimental technique whereby his consciousness can travel in time, but only within the boundaries of his own
life span. In other words, the earliest he can go back to is just after
his birth at the start of the 1950s and he can never go into what
would be his future.

Dr Sam Beckett's consciousness in 'Quantum Leap' randomly
flits about and inhabits a new body of some ordinary person each
week. Everyone in that time period sees the person as they always
were, but Sam knows what is happening and is in contact with the
late 1990s. There is always a reason why he has entered the body
and, almost laughing at the paradox, this is usually to change an
event in a person's life in either a major way (e.g. to prevent them
being killed) or in a minor way (e.g. to help them win a baseball
game). Only when Sam succeeds can he 'leap' to another body. If he
were ever to fail, it is theorized, he would not leap but remain
trapped in the past. There is an underlying theme that the purpose
for all this quantum leaping is dictated by 'God' who is using the
unfortunate scientist a bit like a guardian angel. Wisely that theme
is little pursued in the stories which work on their simplest level as
human fables but with the spice of the quite clever time-travel
mechanism.

There are many researchers, myself included, who are coming
around to the view that consciousness probably is the key to time
travel and that as and when some practical mode is found
'Quantum Leap' may not be all that far removed from the truth.

To see this it is time to leave the theory and the fantasy behind
and enter the realms of what is reported as reality. I have little
doubt, both from my own experiences and the countless investiga-
tions made by myself and other researchers into well-attested
stories from around the world, that there is something fundamentally wrong with our view of time. People are crossing its boundaries in many different ways on a frequent basis.

Time travel is a fact. I suspect that when science finally accepts
this then the day when the first time machine is built – not as far
away as most people think – will be brought considerably nearer.

2

A Sense
of History

IN JANUARY 1983 I faced an unusual problem. A woman from Southport had contacted me to explain that she knew where her husband was. This may not seem particularly weird (in so far as I know he was not a wandering gigolo). The difficulty for me was posed by the fact that his whereabouts were supposed to be unknown. Indeed he was on the run from the police! Any moral dilemma that I faced as a result of this information disappeared when the woman explained further. She was describing an 'empathy' which was less than telepathic, more like an emotional bond. In 'Star Trek – The Next Generation', the character Deanna Troi is meant to be a Betazoid alien with this same ability. As a result she has become the ship's counsellor (i.e. psychiatrist) using her abilities to read the emotions of the 1000 strong crew and guiding their mental well-being.

But the woman from Southport was no character from a science-fiction series and she was experiencing this empathetic relationship with her husband for real. She told me that whenever the bridge between them was present it was like 'liquid surging to fill an empty space' . Her emotional volume built up over many minutes until she was full to the brim. Then she was immediately able to sense his presence and where he was at that point.

My investigator's code of practice did not allow for such an eventuality. Yet I realized that to wander into a police station and 'grass' on this man for what, I gather, was not a particularly serious crime, might have theoretically been the right thing to do, but it would also have been rather foolish. One can just imagine signing a statement explaining that my source of knowledge was the fugitive man's wife who had some strange sort of paranormal bond on a mind-to-mind basis which allowed her to sense what he was up.

I doubt the desk sergeant would have thanked me for that.

This story is relevant because it illustrates something that most of the world's population knows all too well; unless, of course, they happen to be a scientist – in which case they will probably have unlearned it. Human beings are not just biological computers with all of our functions decreed by Newton's limited laws of nature and our brain, as some kind of machine, being the sole arbiter of what is real. We also have something ephemeral and scientifically undetectable, called the mind, which often seems far more important than any mass of cells no matter how complex their genetic encoding proves to be.

Mind is a four-letter word to most physicists, biologists and psychologists. It implies something which they know to be untrue – that consciousness is somehow more than an epiphenomenon generated by the wiring of the circuits in our brain. To them we are no more conscious than a TV set is conscious of the programmes that it displays. We no more have the ability to transcend space and time as a consequence of our mind than does a clock have the ability to turn itself back.

Unfortunately, this rock-hard certainty is hampered by something rather awkward – the overwhelming mass of human experience. This quite clearly dictates that there is such a thing as consciousness, that our mind is by no means just the workings of a time-restricted brain and that as a result many seemingly inexplicable events take place. They are inexplicable only because scientists reject the obvious solution – that the brain is but an instrument which the mind uses to communicate. Strip away that imposition – which intuitively we all know to be refuted by events (although, of course, since intuition is based within the mind, then it, too, must presumably be nonsense!) – what then do we get? A rather dramatic revelation.

As one less-than-blinkered physicist, Arthur Eddington, did say, 'The stuff of the world is mind stuff. ' He was not talking about ESP, but the necessary missing ingredient to make complicated physical experiments and theory work. If you do not introduce consciousness, new science falls apart. Fortunately, more and more of his colleagues are waking up to this truth, allowing cosmologists like Dr Paul Davies, from Adelaide University, to write books about how modern physical science and God are coming together as working concepts. Not so long ago Dr Davies would have probably been burnt at the stake for saying something like that.

This ability to be empathetic does not stay within the boundaries of time. Nor is it trivial in importance. Think about the experience reported by Vaunda Johnson of Arkansas in the USA. All mothers some time or other report how there is often a strong empathetic bond with their children, but generally we think that there is nothing strange about that . . . until, one day, it saves a life.

Mrs Johnson was deep asleep when she was awoken – not by a vision, or a premonition, or anything as concrete as that. She merely had an inner urge to rush from her bedroom into the nursery where her two young daughters slept in bunk beds, one atop the other. She acted swiftly, instinctively, thankfully not stopping to ask herself awkward questions or to allow the rational mind to intrude and point out the bizarre nature of what she was doing.

Vaunda took Karen from the lower bunk, cradled her, without waking the child, and carried the girl into the adjacent bedroom to sleep with her mother. Her other daughter, Yvonne, who was equally unwakening, she left to sleep soundly in the upper bunk.

Minutes after returning to the bedroom, still not having analysed this purely instinctive act, there was a crash and a muffled scream from the nursery. Mrs Johnson rushed in to find that the upper bunk bed had collapsed carrying Yvonne with it. The girl was quite unhurt. But the steel support from the upper bunk was embedded in the pillow, right where Karen's head would have been had her mother not responded to that deep, inexplicable and (science tells us) completely impossible urge to get her out of there.

Now is the suitable time to think about these actions. Because, clearly, there are only two possible ways to answer an experience like this. Science can either try to convince us that it was just a freak coincidence, or we can try to persuade science that it demonstrates the existence of a non time-bound consciousness and the apparently meaningless nature of time that emerges from this discovery. In fact, if you think about it, we do not need experiences such as this one to suggest that mind crosses time as well as space when it forms an intuitive or empathetic bridge.

Take the case of Maureen Blyth. On 7 November 1984 she was in a restaurant in Liskeard, Cornwall, unable to eat her food, feeling in a depressed mood and simply knowing – without knowing how – that her husband was in desperate trouble.

Mrs Blyth's husband is called Chay – the famous round-the-world yachtsman. At that moment when his wife had this acute sensation he was being cast from his catamaran into the freezing waters of the South Atlantic, where he was trapped for 26 hours before rescue. Obviously, this is just one more example, from thousands on record, of that impossibility – the empathetic bond. Maureen Blyth did not know what had happened. She had no premonition before it occurred. Her mind and her husband's mind somehow bridged the thousands of miles between them as if these were irrelevant. Which is exactly what they may well be at the level of our consciousness. However, if we probe more deeply into this experience we see that questions of time do enter into the equation. There is no evidence, here or in other similar cases, that phrases like 'at the same moment' do not mean exactly that. Yet if some sort of communication was passed between husband and wife then how did it travel? More intriguingly, at what speed did it pass?

There have been experiments conducted, usually in secret, on both American and Russian space missions in which some sort of information has been conveyed mind to mind from space to earth. Tens of thousands of miles have proved no hindrance. At outer space type distances even radio messages at the speed of light take a perceptible time. Fast as it is, electromagnetic radiation needs a finite period to travel rather a long way. However, if empathetic rapport really does occur simultaneously, as seems to be suggested by these cases, then it is travelling faster than light and bypassing time just as mysteriously as it is traversing space.

This ability is far more common than most people realize. Indeed, I believe it is a sense we all possess but which most of us censor out. We have what I call a 'doorman' standing guard over the entrance to our rational, deductive brain and filtering these memos from the subconscious mind as they are constantly trying to get through. Partly this is because we have become indoctrinated into accepting 'gut feelings' or 'instincts' as less important, but it is also a natural process to prevent the brain overloading with data. This filter has become a part of us. But beneath the surface the empathetic information is still there much of the time, recognized by those of us who have grown from child to adult without being taught to reject it and instead learning to accept and utilize its potential.

Mrs Jopson of Kent illustrated this effect by describing what happened in April 1973. Her eldest daughter, not then three, was playing games with her when the child suddenly announced that her grandfather was ill. Puzzled, Mrs Jopson asked for more details and the girl kept insisting that her grandfather, then almost 80 km (50 miles) away in London and perfectly fine so far as everyone in the family knew, was in fact very ill.

Later that day the news came through that he had indeed been taken suddenly ill. In fact, as was later established, he suffered a heart attack apparently at the precise moment when his two-year-old grandchild 'tuned in' to that seemingly unknowable fact all those miles away. Of course, at two, she was not yet old enough to have been brainwashed by society into regarding this as impossible.

Any child who has such a thing befall them is best not treated as 'strange' or 'different', because then it is possible that they will continue to have these experiences throughout their life. But it would be all too easy to convey – no doubt subconsciously – the image that these things are 'bad' or make people view you 'oddly', as a result of which the doorman will probably ensure that few, if any, more of them ever get through.

TV actor Bill Waddington (who plays lovable rogue Percy Sugden in the hit soap 'Coronation Street') told me how this ability has aided him.

As a young man in the entertainment front line during World War 2 he had such experiences where he acted like a 'sponge'. If he were amidst a group of soldiers who had just come through a fearful battle he could not only sense their deep despair, to the point that it affected him, but he could draw it out from them and so help them get through the trauma.

More recently, owning thoroughbred horses as he now does in a stables on the Pennine Hills, Bill has had situations where he has had an empathetic rapport with his stock – which, to him, are more than just animals, of course. He could call the stables first thing in the morning, request information on a specific horse which he 'knew' had been ill during the night, and astonish those in charge who had not had the opportunity to tell anybody about this. Questions such as 'how did he know?' are unimportant. It is enough to appreciate that he did know and that, like so many of us, this is an apparently latent ability that is far more widespread than we generally realize.

Rather more specifically it has its advantages if you use it in your job. Renie Wiley does so, as a police detective who graduated in Florida, USA. She has trained her empathetic abilities to sometimes translate into flash pictures which can be used like a jigsaw puzzle to offer clues in crime solving. But it is something she has had all of her life.

As a child Renie sensed the fact that one of her teacher's car tyres was flat – so she warned her. Sadly, the teacher was not enlightened about the things we are discussing in this book and concluded that young Ms Wiley could know this (soon verified) truth only if she had flattened the tyre herself. So she was frogmarched along to see the principal.

In this man's office the empathetic sense came to Renie's rescue. For she 'picked out' from the principal's mind a scene that had occurred the night before. He had been out to dinner with an attractive woman, just as the girl faithfully reported. The man was shaken, partly because she should not have known this by any obvious means. But the principal was also quite willing to offer up an instant pardon, possibly not unconnected with an understandable desire that his wife should not somehow hear the news that Renie had 'tuned into'!

Margaret Rickard* from Lancashire also utilized the ability quite effectively. She told me how one day she was seriously ill in bed and unable to move. Her husband had put a stew on the oven and gone out whilst he let this simmer. But he must have miscalculated the intensity of the heat as the pan rapidly began to burn and Margaret feared disaster was imminent.

As Mrs Rickard lay there beaming out a distress signal to her husband, pleading with him to return, he did exactly that. He explained that he had an 'urge' to turn back and knowing that something was making him do so, but not what, he hurried home, in time to prevent what might have been a catastrophe. This was no premonition. It was a distress flare between lovers.

I suspect that some apparently psychiatric illnesses, e.g. schizophrenia, may at times be better understood as a breakdown of the doorman mechanism. If this natural inhibitor which prevents Psi (pronounced 'sigh' – meaning paranormal sensory information) from getting through to the brain, somehow fails to operate then we may find that a deluge of Psi-emotions are flooding in from our consciousness. By not filtering these out the brain could suffer a sensory overload. It drowns in psychic imagery.

There is a hint of this in the number of schizophrenic patients that I have heard from who seem to have had genuinely puzzling experiences, amidst their more imaginative ones, and who also often find that learning to cope with, and so control, these empathetic responses can sometimes be a better way to overcome the physical illness than constant drug therapy. However, I stress that I would be the last person to suggest that such people should not seek the help of a medically qualified therapist first and foremost.

Audrey Hampton* from Philadelphia, USA, told me of her own experiences after she was diagnosed as suffering from MPD (multiple personality disorder). In this condition the brain is so overloaded that the mind splits into several, unique, sub-personalities, often inventing names for themselves and seeming to be at war with one another inside the patient's head. It can be a frightening phenomenon.

Audrey had always had an empathetic bond with her then teenage son, who had a disorder which made it difficult for him to read or write. Once when she had taken an overdose, he had 'sensed' it and come home in time to help.

During her protracted MPD therapy with a noted hospital and psychiatrist several hundred miles from home, Audrey suddenly plunged into deep despair. She started talking about death and sensations that she was picking up, but both she and her doctor wrote it off as part of the 'exorcism' of her illness. This was on 29 March 1984. Hours later she cut a cord from a hair dryer and tied the flex into a noose around her neck and walked into a shower cubicle in a sort of trance. Luckily she snapped out of this state just in time to prevent herself doing anything drastic.

The next day her husband arrived at the hospital to break some tragic news. Twenty-four hours before, at the time when Audrey first had the empathetic bond with death, her son – then aged 13 – had died in a terrible accident. He had been playing 'escape artist' but a trick involving a noose around his neck had gone fatally wrong.

This concept of Psi information at the level of consciousness being potentially both timeless and spaceless offers a useful way to understand all manner of paranormal experiences. We can draw a chart which shows how to account for virtually all of the seemingly 'different' phenomena that we will investigate throughout this book.

For example, what we call the 'future' or the 'present' in our brain-restricted definition of time must also stretch to encompass what we interpret as the 'past'. At the level of consciousness it will be just as accessible to our mind. The evidence shows this clearly. There are what we can term 'sense of place' experiences where witnesses pick up 'vibrations' from a location which seems soaked in heavy emotion and sensations.

Aughton Moss in south-west Lancashire is such a location. Superficially if you visit you just find cornfields, the town of Ormskirk, scattered farms and the sands of Formby beach. Yet beneath the surface lurks much more than that, if we are to take seriously the views of many witnesses.

In 1644 at this site there was a terrible battle which was decisive in the English Civil War. The royalists were attacked at Ormskirk by a band led by a Scottish vigilante general. Reinforcements were marching to the scene from Liverpool and would have turned the tables had not harried soldiers mistaken their saviours for further enemy troops. Royalists slaughtered royalists in this tragic error and Aughton Moss flowed with blood.

There have been many stories from people, utterly unaware of what has happened in this quiet region, apparently tuning in across time to the emotions of this battle. One man on a walking trip reported how he felt sure there must have been a terrible accident because of the pain and suffering that hit him like a tidal wave as he strolled the country lanes of this area. Others have sensed sounds of battle or horses galloping across their path.

On Formby beach fellow investigator Peter Hough and I have independently investigated three reports, from unrelated sources, of startlingly similar 'poltergeists'. In one typical case a courting couple parked by the sand dunes could not get their car to start, or open the boot (which seemed magnetized shut) as sand flew around them in a frenzy and a yellow glow saturated the area.

There are many other apparent detections of the mood of death, despair and battle-crazed frenzy within this small, rural zone.

Elsewhere around the world similar phenomena occur. There is 'a heavy pall of psychic energy' (as one researcher phrased it) around the now derelict sites of former Nazi concentration camps where unimaginable suffering occurred during World War 2. Even animals will not go near according to some sources.

Whatever its source, this energy, Psi, is decoded through the doorman and into the brain, where it is then apparently interpreted according to the rules of time and space. We impose these. There are two 'modes' by which this information is processed – via feelings and vision. Mode one I call the Psi-emotion and is, I suspect, the root-level phenomenon, in that consciousness detects raw emotions far more easily than anything else. This would therefore immediately explain why emotion-laden phenomena (death, tragedy and battles, and so on) are often the source of a time-shifting experience.

The Psi-emotion manifests as a mood, feeling, presentiment or hunch, according to our terminology. If it seems to come from the past then we have a 'sense of place' experience, where we feel good or bad 'vibes' once trapped there. If it seems contemporary, then we term it an empathetic, or even telepathic, link with someone. If it relates to what our brain decides is the future then it is interpreted as a sort of premonition ahead of time.

At the level of consciousness, where time does not seem to be important (indeed we may eventually find that it does not exist) these distinctions are arbitrary. They become significant only because our current world view decrees the need for them and is dictated by the models of reality that science, rationality and the logical cortex of our brain construct. These seem to need this split between past, present and future for some reason.

In certain people the Psi-emotion goes further. Evidence from research by psychologists suggests these people are those who are gifted at visual creativity (i.e. they are good at art, poetry, or just drawing pictures in their mind). The raw Psi-emotion is transformed into visual form, which is obviously more dramatic and memorable. They will therefore recognize their Psi-level experiences far more often than others and become termed what we would call 'psychics'. This almost certainly means no more than that they have better visualization than others who are also to some extent 'psychic' (and, I suspect, this probably means all of us). But creativity tends to allow for more vivid experiences – which we can call Psi-visions.

If these people 'tune into' something the brain decodes as the past they may have a 'time-slip' or a 'past-life vision'. If the information relates to a more or less contemporary event (e.g. a death or illness of a loved one) the phenomenon manifests as a crisis apparition. If the future is the source of the information, as determined by our brain-shackled view of time, then we may literally see the future before our eyes.

In this way, virtually every form of paranormal experience related to time becomes readily understood by way of the same straightforward process.

At the level of consciousness or mind all time is one and we potentially have access to any part of it. Information – i.e. Psi – enters past the doorman into the brain this way and, more often than not, gets no further than that for all kinds of reasons.

When it is allowed through we then have to translate it into time-related form suitable to the operation of our rational, deductive brain and its time-restricted view of the universe. We most often pick up nothing more than the message's raw essence – i.e. a Psi-emotion. If we are particularly good at visual creativity we may well transform this into a far more powerful Psi-vision. In both cases the distinction between past, present and future is imposed upon the incoming data by our brain seemingly as a means of classification and for the purposes of storage and comprehension.

However, the evidence from real experience appears to suggest, at the universal level of consciousness – that wellspring from which all information is drawn – that time is an artificial construction lacking the rigidity we have sought to impose upon it. In this way the mind can time travel into the past, present and future.

3

In Touch with Yourself?

ONE OF THE STRANGEST couple of hours that I have spent was the day I witnessed my own death. I was 48 at the time, living a hand-to-mouth existence within a wooded area near the village of Alferington (now called Alvington) on the edge of the Forest of Dean. I have never actually been there. Nor, as I write, have I experienced being 48 years of age. When this happened, my name was Mary Reynolds and the year was 1782.

This experience all came about thanks to a remarkable woman called Mary Cherrett. A nurse by profession, she specializes in a technique to help people 'visualize' a past life without artificial aids, such as the somewhat controversial method of hypnosis. In fact we were not even in the same room when we went through this procedure. She was in Surrey and I was 200 miles away in Cheshire on the other end of a telephone!

The episode was intriguing. I expect that it involved a degree of auto-hypnosis, creative visualization and relaxation therapy. But I simply let my mind drift away and – through Mary Cherrett's expert guidance – described what was coming into my head in moving picture form.

The images that I saw were clear and unexpected. If I was imagining, then it was in the same way as one does in the dream state. In effect this was a package tour taken through my own head. The images popped up like magic out of nowhere and I had no way of being sure what they meant, if they meant anything at all. Their source could have been pure fantasy, memos from my unconscious, that wellspring of timelessness we discussed in the last chapter or, perhaps, some real past life that was being unlocked from my psyche. There is no objective way to distinguish between these choices and so I perceived it as merely a fascinating experiment.

I was gaining a first-hand perspective on a subject that is accepted doctrine for half the world's population . . . that we live more than one life.

Reincarnation or past-life study is very popular amongst the new-age community. The writings of Hollywood actress Shirley MacLaine have turned many into believers, but science is predictably (and not unreasonably) sceptical of it all, preferring to seek other answers. There are real difficulties in accepting what the evidence dictates at face value. The concept most often applied is that our 'soul' is eternal and if we expect to live on after we die, then where were we before our birth? The answer, for millions, is that we were probably somebody else.

Of course, trapped in this time-blinkered prison cell of a brain we cannot conceive of any existence before our birth just as eternity leaves us floundering. Yet researchers suspect that images from those times trickle into the subconscious and sneak past the doorman, occasionally intruding into dreams or as flash pictures we do not recognize. The easy way out is to assume we have invented these images – and the mind clearly can invent non-realities with ease so that is far from impossible. The other approach is to see these images as a hidden memory, vestiges of a previous you before the inner you became the present you! – if you see what I mean!

However, if consciousness is timeless then we need not have to believe that we are the person to whom the past-life memory relates. If, as we saw in the last chapter, you can stand in a field where centuries before a battle was fought and detect the emotions, perhaps hear sounds or even see 'ghosts' from that time, then this does not imply that you were a part of the battle. Similarly we might merely tune in to a person who lived in the past, absorb images from his or her life and 'synchronize' with them, thanks to some mental trick like hypnosis or Mary Cherrett's guided imagination.

Armed with this reassurance that one does not have to believe in reincarnation to have a past life, I let Mary take my mind on a journey. She held the leash and my consciousness followed in strict obedience. First she took me back to scenes in my childhood. She did this, as I think I could have predicted, by asking me to home in on my feelings and flesh out images around these. That makes perfect sense, because, as we have already seen, the raw Psi-emotion is the thing that most often gets through the barricades of time imposed by the rational brain.

My first stop was aged four, where I described feelings and images that later were verified. Then I experienced my own birth. That was really eerie.

The process resembled a word association test. Once I realized that I had no need to 'remember' (everything was being tape recorded) I played the game. I just spoke whatever 'felt right', freely tumbling out of my inner self in a stream that largely passed me by. I was just drifting on a river enjoying the scenery that was leisurely floating alongside.

I described my own birth process, the room into which I was born and the apparent struggle I put up to stay out of the birth canal (saying that I did not want to come out but was forced because I had a job to do!). Later I realized that as a young child I had had many frightening dreams of being stuck head first in a narrow pipe and feeling that I was suffocating. Mary Cherrett told me that these (now long forgotten) nightmares are common for children who have difficult births.

I had no way of knowing whether any of this was true. But I asked my mother. She confirmed something I did not consciously know, although it is perfectly possible that I had heard it said at any point during my life and simply forgotten about that moment. My birth was indeed a long and difficult process. Equally, she heard my account of the room into which I was born (at an old hospital in the Rossendale valley) and affirmed that it was not unlike her memory of the real thing. So – if I could be accurate about an account of something that I clearly did experience, but which virtually all traditional scientists say, as a still unborn conscious entity, I cannot have 'experienced' – was it possible that my 'life' as Mary Reynolds was also a reality?

Certainly it was not without detail. I described my birth and marriage at Church Lea in Bristol – a city I had visited only once (for about four hours) some years before. I gave graphic accounts of how my husband Frank disowned me, of his attitude, seeing a clear picture of him sticking a hay fork in the ground and saying that after seven years of marriage that was all he had. I then went on to describe how I lived in the woods, cooking stews in a big pot and trading with passing stage coaches (the original motorway service franchise!). Finally, I visualized my death from a nameless wasting disease which the 'apothecary' could do nothing to help. I died alone, lonely and puzzled.

This was hardly the kind of romantic nonsense you might imagine someone would dream up. It was routine. Little happened and in the end it was a slow, painful and far from spectacular demise. But was it fantasy or memory?

I contacted Ian Wilson, who is one of the world's leading researchers into past-life accounts. His book *Mind out of Time* is probably the most important yet written on the theme. His approach was, I knew, very sceptical. He was not going to tell me I had really lived before just to make me happy.

He certainly did not tell me that. He pointed out that Mary Reynolds was a noted real-life 'Rip Van Winkle' who went into a coma in Pennsylvania one day in 1811 and awoke with a multiple personality. Chances are I had read about this case somewhere, but I had no conscious memory of it. That had to be added to one fact that worried me (indeed I fought against it when I first spoke my name in this past life). Mary was the first name of my guide and Reynolds was not all that different a surname from my own. It just seemed suspicious to my rational mind. Ian Wilson quite reasonably cautioned: 'Is your subconscious trying to tell you something?'

Of course, it may well have been. Just exactly what was the importance of the fact that Mary Reynolds' cottage fitted a frequent flash image I have had of such a place? In the past I had assumed it was one that I had imagined as a home I would like to own one day, or an image from a story book read as a child. But did that image trigger the past-life memory, or were vestiges of the past life in my subconscious as the origin of these flash images of that cottage? It was impossible to say, just as it was impossible to argue whether I have a peculiar affinity towards Bristol – despite having no familiarity with it – because I once lived there. Or was such a fondness the reason why my mind located a past life in that city?

One thing was surprising, Ian Wilson had no intention of searching parish records to see if in 1759 Mary and Frank Reynolds were married in Church Lea, or if indeed there was a place so named. Even if it were traced it would prove nothing, because I might have seen reference to the fact in some long-forgotten obscure location, which lodged in my mind. What he could tell me, however, was that Alvington was a real place, located just where I (or rather where Mary) had said it was. I looked it up on a map and sure enough the little dot was sitting on the edge of the forest.

Ian Wilson explained that he felt certain: 'The information may have been from some story you read ages ago, in which you never even consciously noticed the name, but it can none the less have stuck. ' He called finding this source a 'needle in a haystack' and probably unverifiable anyhow.

My own forays into the world of reincarnation illustrate all of the problems which the subject creates. They have been wrestled with for decades by serious researchers. What I went through is very typical. The last thing that occurs is that people regress to lives as princesses in ancient Atlantis or swashbuckling pirates. Yet, if this is pure fantasy, would that not be exactly what we should expect? To find a tirade of mundane, fairly dull, routine lives with often early and unspectacular deaths is puzzling.

In a series of hypnosis-based experiments with past-life therapist, Joe Keeton from the Wirral, which I carried out for the BBC in 1986, we led a previously cautious producer back to a past experience as Jill Leadenoak. We could get almost nothing out of the normally vivacious woman in this state. 'She' was a pig farmer's daughter who did not even know the year. She would happily talk to us about the names of her pigs, about the disgusting (but – as we discovered – historically accurate) methods of brewing cider – but, despite much effort, little else.

Yet 'Jill' spoke in a delicious accent, far removed from the cultured tones of the woman who 'became' her. She was embarrassed when first hearing the tapes, but this soon turned into astonishment. Despite most of us (including the unhypnotized BBC producer) thinking that the accent was probably from Somerset, it turned out to be from the Hereford area. With so little information it was impossible to find whether Jill Leadenoak, pig farmer's daughter, ever lived near the 'Green Wood'. But if this was a subconscious fantasy (and it was definitely not a conscious one), then it was constructed over many weeks and hours of hypnosis sessions and why was it so blindingly dull? Our subject was a brilliant, well-educated, highly creative young woman. She could have conjured up a far more imaginative story with ease and outside of any altered state of consciousness.

Dental surgeon Leonard Wilder from London uses a similar technique. An insurance agent I once met told me how he had been hypnotized by Dr Wilder and vividly recalled being shot down in flames during World War 1 – when he was a German pilot, babbling hysterically in that tongue (which he did not speak) and screaming 'Kaput!' as he crashed to earth.

Journalist Gabrielle Donnelly was led by Dr Wilder back through her childhood and eventually into the past in an experiment rather like my own. She had vivid pictures of a life as Alice Baines in 1672 and a typically mundane existence, where there was little to describe except the clothing that she and her fiancé wore. As the (twentieth century) woman said of this: 'Was it regression, or was it simply an image that had floated into my mind from who knows where? I didn't know . . . but my mind fixed on the idea, worried it, developed it. . .' I can confirm that this is how it works.

Leonard Wilder thinks that the process of retained but forgotten memories which Ian Wilson champions (known to psychologists as cryptomnesia) is responsible for some past-life data. But he feels that it cannot explain everything he has uncovered, as at times information is too specific, real emotions are lived through, actions expressed. He played some of his tapes to a friend, award-winning actress Diana Rigg, who told him that if the subjects of these sessions were acting they were good.

I have to agree. I have records of taped regressions of soldiers reliving terrible injuries, of a man drowning, of another in a drunken stupor and so on. The people who are 'reliving' these things are not professionals, nor even practicising amateur actors. Yet, if they could harness this talent (if it eventually proves to be no more than that) then they would give Meryl Streep a run for her money in the battle for the Oscars. At the very least this process of tuning into past-life memories opens access to a remarkable latent ability. But it is conceivably more than that and, instead, might lead us into uncharted territory of mind and time.

Joe Keeton is unsure what explanation to apply. But he does have two interesting observations. His patients never seem to cross racial barriers. If they are British today, their past lives appear to be British also. Other researchers have differing results. This is curious, some might even say suspicious, implying the phenomenon may be a form of imagination at work. Yet, almost paradoxically, Keeton finds that if you progress back in time much beyond the fifteenth century you cannot communicate with the subject. They do not seem to comprehend your questions. This again contrasts with other researchers, who report communicating with past lives in prehistoric times when language was barely developed and far removed from modern English!

The problem always must be that there is a bridge between today and yesterday. That bridge is the mind of the subject of the experiment. They will obviously understand the questions put to them and, if really accessing images from somewhere in history, can describe them even if they cannot 'communicate' verbally with the person in that dim, distant past. In effect, via this process, we are turning human beings into time telephones, allowing us to talk to people from long ago by utilizing the vocal chords of the subject. It hardly matters whether the phenomenon is a mental trick or real time travel. It is a fascinating experience either way.

Indeed, even those doctors who cannot accept the past-life source of these stories often realize the psychological benefits of the process. Dr Gerald Edelstein of Herrick Memorial Hospital in Berkeley, California, is a case in point. He has found 'past lives' cropping up in his work to resolve deep phobias in people's lives. When he uses hypnotherapy to try to cure them they do sometimes spontaneously regress to a prior existence where the source of the phobia is seen to emerge. Simply denying that this really is proof that we have lived before does not diminish the potential value of the catharsis that this experience can bring to the patient. In other words, who cares how it works, so long as it cures.

Dr Edelstein cited an example where a woman traced a phobia of childbirth back to a scene in 1793. Her (past-life) mother had bled to death after giving birth within a remote log cabin where there was no medical aid. Dr Edelstein tried to show that this was a screen memory for something that had really occurred in the woman's own earlier childhood, but he failed. He could not believe she had really lived before as the daughter of a poor trapper, but the regression to this scene (and others) quickly eased the phobia towards a full recovery.

There are countless other cases from the files of many psychotherapists where traumas and phobias have been removed in similar ways. A fear of red, stemming from a parent being stabbed to death in a bloodied frenzy. A fear of water, coming from an accidental drowning. Whatever the source of these magical cures seems barely important in the face of real clinical improvement that they seem to bring about.

American Joseph Vitale described how he struggled with his weight, at over 660 kg (21 stone) and was unable to find any effective diet. But he was led to a past life by a process similar to Mary Cherrett's guided creativity. Here he saw himself as a monk halfway up a mountain in the Himalayas. The winter was bitter and there was no food. He gradually starved to death. He decided he was compensating for that life by over-eating now. As soon as this was accepted, the phobia (which had not even been recognized as a phobia) lessened and within nine months Joseph lost 200 kg (6¼ stone) with ease.

Californian researcher Loyd Auerbach is worried that 'some people are using their past lives as escape mechanisms' and this may place them in 'psychological jeopardy', despite cases with positive outcomes. He is unsure whether evidence supports reincarnation, a position agreed by most researchers. The problem of cryptomnesia in each case is the real bugbear.

There are reports, uncovered by Ian Wilson and Melvyn Harris, for example, where a graphic past-life 'memory' has been indisputably traced to stem from something the hypnotized person read years before. In one classic instance the life of a woman who saw herself as Livonia in Roman times was proven to emerge from historical novels. In fact some of the characters in the past life were ones invented by the novelist! As Wilson insists, if it can be shown in one case then it is potentially demonstrable in all others.

One of the recordings I made with Joe Keeton was of our sessions with a Reading man, Ray Bryant, who was reliving the life of Reuben Stafford, a man with a (passable) Lancashire accent who fought in the Crimean War. His file has actually been uncovered by Fulwood barracks in Preston. This man really existed and many of the details offered under hypnosis by Bryant fit his actual life. Ray Bryant also described his own death by suicide, unaware that as he was doing so Joe Keeton had traced and secured the death certificate of this man more than a century ago! That form gave details which agreed with the death scene that the hypnotized Ray Bryant produced.

The question which nobody can answer here is whether the discovery of these records makes reincarnation more, or less, tenable. It seems to make it more tenable because Ray Bryant insists, and I believe him, that he has never consciously come across any of these details before. But how can we be certain that he never saw them in some obscure source he has now forgotten?

That is the dilemma. Establish evidence to support a past life and the sceptics can immediately (and rightly) argue that the person reliving the past life might have stumbled across that evidence too. Fail to trace any details of the life and you can never prove it ever happened and so is more than fantasy. Seemingly it is a no-win situation.

Indeed, even Ray Bryant himself, despite his remarkable regressions, draws back from saying that he ever 'was' Reuben Stafford. Only that he seems to somehow 'identify' with that man from the past.

It is the regression-hypnosis state that seems to worry cautious scientists more than anything else about past-life studies. This artificially induced altered state of consciousness can facilitate both imagination and recall in just about equal proportions. Even experts struggle to tell the difference.

Dr Michael Heap, secretary of the British Society for Clinical Hypnosis, argues with both Melvyn Harris and Ian Wilson about the extent that cryptomnesia plays in all of this, but he also does not believe in reincarnation. He suspects that things may be a little more complex.

Heap draws a parallel with our use of role-play fantasy as a child or at school. He also points out the importance of 'priming' by the hypnotist which ensures, most of the time, that the witness expects to be taken to a past life during the coming session. In the hypnotic state suggestibility is very strong and images can be created to fulfil this expectation. He points out that experiments have shown how a person can be asked to 'pretend' to relive a past life and the outcome is not at all dissimilar to supposedly real ones emerging under regression hypnosis. As such, when we know that some cases are undeniably nothing but role-play fantasies, why not all of them – with most of these arriving unconsciously?

However, it is misleading to think that the evidence derives only from hypnotic regression. Indeed my 'life' emerged in a non-hypnotic state and there are plenty of others which come from similar situations. These usually involve some degree of shift in consciousness, which may be necessary to somehow trick the doorman into allowing access to data from that timeless void that we call consciousness. As a result we have reports of past lives coming in reverie, daydream flashbacks, during Yogic relaxation and dreams.

Paul Catlow from Norwich told me of his extensive, recurrent dreams of his death as a soldier called Karl Schumann from Hamburg on 23 October 1944. He has no idea if such a person did live and die, but has verified some of the detail, such as the unusual colour of the uniform worn, being one that we would not expect from watching TV series or movies about the war.

Of these very realistic dreams Paul says he would be delighted if he could only accept them as 'an elaborate self-hoax'. But he cannot do so. Yet, he also wonders why, if he really 'was' a German in a life terminated in the very recent past, he finds the German language so impenetrable.

This really is a tricky business, juggling contradictory evidence and weighing diverse possibilities together.

By far the most remarkable research carried out so far is that conducted since 1960 by Dr Ian Stevenson of the University of Virginia School of Psychiatry. His erudite papers, containing millions of words, have researched hundreds of cases from all over the world and focus particularly on two things that make them so important. They are stories told by young children. And they are spontaneous memories, which tend to begin as soon as the child can speak and then rapidly fade around the age of seven or eight.

Although many of these cases come from countries such as India, where reincarnation is a part of everyday belief and religion, there are reports from areas where that is not true. Indeed an increasing number come from places like Britain and America, now people are alert to the need to look.

I know one British mother whose child began to describe his 'death' as an Indian brave in the American west. She wrote it off as role-play fantasy, until she noticed the persistence, the total lack of any self glory and its simple, almost poignant terms. The child mentioned it in passing once, as if it were the most natural thing in the world. He was five. The family reacted oddly, evidently indicating this was considered 'untrue' and he never mentioned it again, until asked. After two years, when asked again, he could not remember ever having thought about it.

There are suspicions that, if we were more open to this possibility (and brave enough to ask our children seemingly incredible questions about their memories 'before birth') that we might find many more such examples.

Ian Stevenson's research is intriguing. But it has not proven reincarnation. He only refers to it as 'suggestive of' that possibility and indeed there are those who think that his work may disprove the reality of past lives altogether. They point out that in his findings the claimed past life almost always is in a nearby village where relatives of the former deceased still live and often have more money than that of the family of the child who is now claiming the past life. A switch from male to female, from one life to the next, does happen, but it is relatively uncommon and seems only to be reported in cultures where such a possibility is accepted by religious doctrine.

These all sound rather suspicious features. But on the other hand there are cases where extraordinarily detailed 'memories' seem to be retained by the child, e.g. the names of pets once owned, total memory of the insides of the house once lived in, and so forth. There are even examples where a death wound from the previous body seems to relate to a birth mark in an identical spot in the newly reincarnated person.

An interesting result which emerges from past-life studies, regardless of country or culture of origin, is that the closer to today one searches for past lives the more frequently they seem to be occurring. Any individual who is probed to uncover their past lives in some way will almost always come up with three or more (sometimes many more) if enough time is devoted to the search. In my case, we found two others in the limited time available (one in France centuries ago, another which only ended 17 years before I was really born).

In general the thousands of past lives now collected allow for patterns and trends to be plotted from them, regardless of nations, researchers and cultures. The information proves fascinating. It may refer to periods in history scattered like this: 1150, 1500, 1750, 1810, 1900, 1935 etc. As the centuries pass the lives squeeze closer together, but they never overlap with one another. No life starts until after a previous one has ended.

This makes definite sense, because if there are a finite number of souls to incarnate on earth then these would need to be returning more and more frequently to account for the rapid population explosion. How can a worldwide fantasy show exactly this kind of mathematical logic?

However, we are practically at saturation level now (i.e. the gap between lives cannot get much closer than it tends to be). Therefore, if the world's population continues to escalate at its presently horrendous rate over perhaps the next 50 years or so, then this mathematical logic will soon break down and strong evidence against reincarnation will emerge from this failure. If reincarnation is real, one might predict that world population must tail off, or at least level out, very soon.

In 1983, the research group ASSAP (Association for the Scientific Study of Anomalous Phenomena) began a long-term study of reincarnation to compile statistical data.

In early 1993, reporting on the first decade of this research, chemist Dr Hugh Pincott told how it was still not possible to distinguish between rival theories such as cryptomnesia, super-ESP (i.e. the subject picks up the knowledge to create the past life by tuning in to the vast store of information in the universe) or real reincarnation. Another theory is memory inheritance, i.e. data passed on in the same way as genes are inherited within a family lineage. But this often fails when a person in the past dies childless. If they do so they obviously cannot convey memories via genes.

Dr Pincott cited two impressive cases. A man relived life in the eleventh century as a farmer called 'Cerdic' from Kent. He described fighting in the Battle of Hastings in 1066. But he did this only after he begun to doodle during boring meetings. In moments of reverie a man without prior artistic prowess was creating accomplished pictures of life in Anglo-Saxon times which, Pincott says, were verified as accurate, then explored under hypnosis.

In another extraordinary case, which might come as close as we can get to proving reincarnation, a woman relived her life in the sixteenth century. One room in her former house was said to have been converted into a chapel. Researchers eventually traced this Hampshire building, which stood partly intact. The owners were unaware of details but did know that one room had indeed once been a chapel. Using information gleaned from the hypnosis sessions, a hidden cupboard was accessed. The occupiers of the house were completely oblivious to its existence. The method of opening this storage space, as had been described by the woman from 400 years ago, proved to be accurate. Inside, again just as was described, were found authenticated deeds of the house dating from the period of the woman's remarkable 'past life'.

Accepting cryptomnesia in a case such as this, whilst it is not impossible, is stretched to the limits of what we might regard as plausible. This is merely one of several extraordinarily well-attested cases of apparent reincarnation which have come to attention in recent years.

In the USA there is the remarkable story of a group of 35 people from California who have been gathered together by therapist Dr Margie Reider. They have gravitated to her, some by their own volition, others at the urging of people they know. Many are part of one community today, although some do not know one another and only a few are actually related. Only three claim to have ever visited Virginia, thousands of miles away across the continent of North America. Yet under hypnosis the 35 have all described to Dr Reider interlocking past lives in which they lived in the small town of Millboro during the Civil War. They have been able to sketch roads and houses that match this town as it was in the 1860s.

There is a touch of Hollywood about this case which makes one instinctively wary. But Dr Reider appears to be arguing quite sensibly that this is not proof of reincarnation, but rather of some sort of collective unconscious into which we can all somehow tap. Equally, there is no evidence that I am aware of to suggest that the subjects of this amazing multiple past-life saga have any motive but to describe honestly what they have experienced.

In 1993 a case finally emerged into the public eye which may well be the ultimate story of a past life. If you can accept it as an honest portrayal of real memories and the long search to verify these, then it really precludes any possibility of cryptomnesia. It seems difficult under those circumstances not to regard it as strong evidence for a real past life.

The source is Jenny Cockell, a Northamptonshire woman, born in 1953, and a well-educated mother with a good career as a chiropodist. From early on in her life she had recurrent dreams of a village, which she knew to be in Ireland at around the turn of the century. She was able to sketch the layout of the main buildings and roads, describe her own house and pick up various flash images of a wooden jetty, a dog and her children.

As the years went by Jenny Cockell clung to these 'fantasies' as an escape from tensions within her own unhappy childhood. She thinks other children probably have such recall but lose a grip of it as they grow up.

Eventually, happily married and having raised her own children, she decided to explore her fantasy more carefully, although she had really accepted it as an actual memory of a prior existence. She knew the location was Malahide, just north of Dublin in Eire. Getting detailed maps of the village was difficult, but when she did she found that her sketches seemed to match the standing buildings (such as the railway station) and streets.

Jenny did attempt regression hypnosis, but only after much of the data was already obtained. Whilst it added a few images it was not as helpful as might have been hoped. Jenny had long known that the woman from the past was called Mary, but not what her second name might be. Finally, she visited Malahide and began a remarkable quest. She set out to find her eight children from a previous life!

After much effort the cottage was traced and it was discovered that a woman called Mary Sutton had a life which matched details Jenny had recalled. However, her existence was little recorded and some facts remain unverified, whilst others came from talking with elderly local residents. Mary had died in the early 1930s and her eight children were split up and sent to various orphanages. Jenny Cockell believes that this emotional bond and her guilt at having caused them such grief and separation might have been what bridged two lifetimes and allowed her to retain memories as she did.

At last, Jenny Cockell traced all but one of her children (the other was believed to be in Australia and some had died). A remarkable reunion was arranged, taking several of them back to the ruined site of their cottage in Ireland. Here they swapped memories with Jenny in an astonishing manner, as if the years did not matter.

One of the most convincing aspects of this was the little, personal stories that they could share. Jenny (as Mary) had recalled how her son Sonny, age 11, had trapped a hare alive. She was able to show him where this was and accurately describe what happened. A story such as this would never be recorded anywhere and be accessible in the usual way via cryptomnesia. Many may feel that only someone who was really there in 1930 (23 years before Jenny Cockell was born) could have seen it so precisely.

But, of course, doubts will always linger however strong the case. As an example of allegedly real time travel few stories can compare with the moment when Mary Sutton's eldest child, Sonny – at the age of 74 – first met his 'mother' – who was now 35 years younger than he was!

However, judgment of this case may be affected by something Jenny Cockell told me in late 1993. She also has visions of other lifetimes and is now exploring one in particular. The problem is that this occurs (or will occur) in the twenty-first century and involves children not yet born but whose souls may be incarnate (as other people) somewhere on earth right now!

4

Images from Yesterday

DO YOU BELIEVE IN GHOSTS? Most of us are asked this question at some point or another. According to several recent opinion-poll surveys the answer given by almost half the population is a firm 'yes'. More than one in ten actually believe that they have seen one.

This is not a book of ghost stories. You can find plenty on the shelves if you wish to read further so I will not duplicate that work. But ghosts do, of course, have something to tell us about the subject of time travel. In fact, I believe that they offer us a possible way towards the design of the first practical device that will be able to see through time.

The closest that I have come to seeing a ghost was in June 1986. I met a medium in Wythenshawe, Manchester, to conduct an experiment into precognition. With me was a sceptical psychologist from Manchester University to check against any excesses. The psychic told me that I was being 'protected in my dreams' by an elderly woman. She described her clearly and even spoke of her carrying a Sunday school banner. This instantly connected in my mind with my maternal grandmother, who had several strange experiences of an empathetic and precognitive nature and whose interest in the field began my own fascination as a child. She had died in January 1971 and, whilst I had never given it any real thought, the idea of her watching over my interest and research seemed oddly credible.

The problem here is why this medium saw my grandmother as an old, frail lady in the way that I remember her from my early childhood? I must admit that it occurred to me that anyone gifted at ESP reading an image from my mind could have fished this out from my head. It just seemed vaguely disquieting that she saw the gran that I knew so well, not what my gran's own mental image of herself probably was (i.e. a younger, fitter person).

50

Of course, this becomes a concern only when we assume that a ghost is a living entity, returning from the afterlife to haunt the material world. Much of the evidence for apparitional phenomena does not support that concept too well. Rather, it is more suggestive of mental constructions being built by the witness from the information that has snuck past the doorman and entered their mind. It has come from that sea of timeless consciousness.

The Wythenshawe medium knew that the image she saw was not really present, but was inside her head. Had this been a very vivid impression, or had she been especially gifted at visual creativity for example, then more than likely that image could have been projected into the room where we sat. Then she would have 'seen' my grandmother standing in front of her.

This can occur. On one cold winter's morning, I woke early to see my boyfriend leave to go to work 30 miles away. He had just stayed the weekend at my home in Irlam, Lancashire. Given the state of the roads I was not happy about him riding his powerful motorcycle that day, but knew he had to go. Minutes after he left I heard him return and I got up from the settee to greet him, thankful that he had decided not to risk the weather. But as I did so I suddenly 'awoke' to reality. I had not got up from the settee after all and he had not returned. I was still sitting there, in front of the blazing fire. I must have fallen asleep after he had left for work and experienced what is usually called a 'false awakening'. This is an incredibly realistic altered state of consciousness, akin to a lucid dream. It is all sensory and very difficult to distinguish from reality. But it is a hallucination.

In false awakenings, people have been known to experience waking, getting dressed, travelling to work and then suddenly finding they are still in bed – in fact they have never left their bed. They had merely hallucinated that they had. It is usually possible to tell that a false awakening is a hallucination, because you wake up and, as in my case, no boyfriend is present and you can later establish (by asking him) that he had never returned home. However, what if in this state you were to see your dead husband, or an apparition of a person you did not recognize? How could you ever know this was not real? There would be no way to ask them if they were in the room. The fact that they vanished when you 'came to' would mean nothing. It would just make the experience more strange. Ghosts do disappear suddenly.

From my false-awakening experience I would have to say that had I seen my dead grandmother, rather than my boyfriend, in those same circumstances, I would probably now be telling you with conviction I had seen her ghost. All of which may seem to suggest that I am being sceptical, even cynical. But I am not. When I 'saw' my boyfriend the reason why I did so is fairly obvious. It was wish fulfilment. I was concerned and did not want him to leave. But the mechanics by which I created this all-sensory experience of his return remain baffling. If I were to conjure up an image of a dead person in this same way, it would still be a hallucination. That person's spirit would not have returned to earth. But there would have to be a reason why my mind went to so much trouble to create this vivid imagery. Someone recently bereaved might also have a wish-fulfilment motivation. But that is not the answer to all, or even most, ghostly experiences. There are stranger reasons, as in cases such as the following.

Mary Black from Clwyd contacted me to try to understand the vivid dreams/hallucinations she has had throughout her life. Many of these involved the absorption of what we might call emotional energy – as if she picked up the ripples across time sent out by a powerfully emotive event. There was a kind of transfer between physical energy and emotional response. Afterwards she referred to feeling 'drained'. Many mediums speak of the same symptoms after a heavy session, and I think it is because they soak up the emotional energies around them. This is rather like what Bill Waddington meant when he described acting as a psychic sponge. Mary Black told me, as he did, of feeling an 'awful sense of misery' after these ghost experiences. Interestingly, she also reports frequent problems with electrical equipment. If she goes near a calculator or electric clock at these times they tend to go haywire.

In physics there is something called the law of conservation of energy. This says that the energy in a system must remain constant, but it can change from one form to another. Switch on a light bulb and the electrical energy is transferred into heat, which warms a filament. It then glows and gives off light. The combined energy of all of this system remains unaltered. It just manifests in different ways (electricity, heat, and so on).

I suspect that emotional energy has to come into the equation when we deal with ESP or time-travel events.

Mary Black reports how in times of this energy drain she sees apparitions. What is more she undergoes a curious sensation when time stands still. One summer's day Mary was overcome by these things as she stood in the kitchen, her mind idling (a common starting point). Suddenly she saw a man walk through the door, cross the kitchen and leave. He was solid, real and she assumed it was her father who was out in the garden at the time.

Moments later her mother asked her to call her father indoors and Mary pointed out he had just been inside and then left again. This was denied and, indeed, her father had not entered the house. Then Mary realized that she knew the figure seen was not her father, but had talked herself into that idea. She described the man who had been walking only inches in front of her. His clothing was very distinctive. Enquiries revealed that others had seen the figure in the house in the past. He was a man who had been employed as a servant and had lived there for many years and died long before.

What happened here? Probably, Mary had a hallucinatory experience much like my own – seeing a figure beside her that was not there. But clearly this was more than simple imagination, for she saw a real person whom she did not know. She had somehow or other tuned into the past. It was as if her mind was a video recorder and it was replaying scenes from 60 years before.

Actress Pat Phoenix, who played Elsie Tanner in the TV series 'Coronation Street', had a similar experience at her home in Sale, Cheshire. She and several other visitors across the years observed a figure carrying something in her hands. She wore her hair in an old-fashioned bun and had a long grey dress. The entity was completely solid, often mistaken for a real intruder, and appeared all over the house and even in the garden. But she always walked like a robot, doing the same thing again and again and never reacting in any way, no matter how many people were present in the house. Frequently, it was noticed that she would vanish when someone watching her turned their gaze away for just a moment. It was as if this slight alteration in their state of consciousness broke a tenuous lock that was holding the figure within the room. Once broken, the 'signal' disappeared.

In the last century the building had been owned by Madame Mueller, an actress who fell upon hard times. It is generally believed that she is/was the apparition repeatedly seen here. But, if so, then one must assume that she was not present in any spiritual sense within the house. Everything about this experience suggests a form of video replay.

There are many other impressive ghost cases where this idea is strongly reinforced. A remarkable one concerns plumber Harry Martindale who in 1951 was fixing pipes in the cellar of a very old building in York when he saw a horse followed by several weary Roman foot soldiers pass through one wall, cross the floor and head off out through the opposite stone edifice. It was verified that a Roman parade route passed through this way.

That such things are video projections is stressed by cases where the floor level of a building has been raised across the years, or if outside, for example, pavements are now much higher than they formerly were. It is not uncommon for the feet of the spectre to be invisible beneath the ground, as if the image is still walking along the level that existed on the day when this scene was somehow 'fixed' into the fabric of time. The whole thing is akin to how a moment in history can be frozen onto film by a video camera. It then exists forever, trapped as an electrical signal on magnetic tape. If you have the right equipment to unscramble the signal and turn it back into a moving picture you can project this onto a TV screen, fifty, a hundred, or a million years from now and effectively 'see' the past replayed. Given our own technological ability to do this, the idea of time replays from the past occurring within nature itself via some as yet undiscovered means does not look beyond possibility.

The idea of an entire platoon of soldiers and a horse trooping through eternity might put anyone off time travel. Such cases strongly imply that a video signal is being held in a sort of loop in time at this location waiting to be decoded into a visual image. An excellent episode of 'Star Trek – The Next Generation' featured in the 1992 series explored this theory unintentionally, but very cleverly. It developed from the concept of *déjà vu*, the feeling most of us get on a regular basis that a moment they are experiencing has (as these French words literally translate) been 'already seen'. We are vaguely aware of having witnessed it somewhere before.

Usually, the sensation is intangible and transitory and we cannot say whilst it is happening what real events will follow. But afterwards we are sure we had somehow gone through the entire (normally trivial) scene twice. There are many theories designed by psychologists to try to explain this remarkably commonplace phenomenon, but in truth it remains a puzzle. The traditional view most often proposed is that it is a glitch in the recording process by which memories are stored in the brain.

Perhaps when we lay down a new memory it traces a similar neural path within our circuitry to one already established, possibly years earlier. This familiarity registers as the vague *déjà vu* sensation. Another idea suggests there may literally be a brief time lag (a few milliseconds) transferring data from the sense organs to the brain itself. We temporarily misinterpret this, when the data registers in our memory, as a feeling that we had already experienced it all before. We had indeed done so, of course – milliseconds previously. But we cannot place when this occurred.

These ideas are not fully satisfactory and, if we are willing to entertain more paranormal options, these might be preferred. If we do dream of trivial future events as often as the evidence suggests, but forget 99 per cent of this imagery, then maybe when we do experience a reality that we have already dreamt some time before, those barely recorded memory traces can be activated from the subconscious mind. Would this then manifest as *déjà vu*?

Occasionally, as events unfold, I have had a vague recognition that I recorded these images in a dream experiment and can later verify that I had indeed done so. But this does not feel quite the same as the eerieness of *déjà vu*. However, *déjà vu* may occur only when the dream memory trace is very faint and irretrievable or never recorded. The act of writing down a dream during a time experiment will solidify this memory, which will fade rather less than other dream memories will do. We ought not to expect to produce the same sensations from differing depths of memory even if in all these cases a real event was foreseen in a dream but then forgotten about.

Other serious suggestions regarding *déjà vu* revolve around the nature of time itself. If there are multiple possible outcomes to every sub-atomic reaction, as quantum physics suggests there may be, *déjà vu* could involve a transfer of information from one closely parallel reality to our own 'world' where a similar, but not exactly identical, event has just occurred.

As you see, *déjà vu* is an important, fascinating but complex problem.

In the Star Trek episode, 'Cause and Effect', the viewers are mystified in a clever way. They see a short sequence in the life of the *Enterprise*, as it enters a new region of space. Some trivial (plus some not so trivial) events unfold. Then, suddenly, it is as if the broadcasting channel has made an error and loaded the same tape back again when 'part two' begins. The episode seems to return to the start and the same scenes are played over again. When this happens a third time, then a fourth, you realize something is going on within the story-line and begin to understand that there is a space-time warp where one short period of life is being experienced again and again. Aside from their vague feelings of *déjà vu* the Star Trek crew do not know what is happening, because when the loop begins again their memories of what is to come are removed. As such, reality is being replayed constantly in the sensory experience of these human beings, and potentially forever.

One can imagine in reality how ghosts may be the 'video replay' of an event from history trapped like this inside a time loop and cycling through a series of trivial experiences, e.g. a woman walking down some steps, foot soldiers on part of a route march, etc. This scene may be occurring (and recurring) over and over, from the point of view of the trapped energy field. We ourselves must 'tune in' our consciousness specifically to the correct channel in order to observe this baffling sequence of events.

It is rather like switching on Star Trek's 'Cause and Effect' part of the way through and watching this repeated sequence of scenes unfolding before our eyes. If we then switch channels, to return ten minutes later, we will find the same scenes are still being screened on our TV in the same order and in a sense witness a time-loop experience. Reality, of course, is not fiction, but video-replay ghosts suggest that it may be happening something like this Star Trek scene.

There are some grounds for believing that the building itself has a part to play in trapping this signal across aeons of time. Areas where repetitive figures are seen doing the same thing over and over are usually said to be 'haunted'. The effects appear to last for centuries – or, if the Martindale case from York is representative, even for millenia.

It has been noticed that sandstone, millstone grit or rock with a heavy quartz content is particularly prone to trapping a haunt. Many old buildings, or newer ones in more rural areas, are built from such materials.

There are also a number of cases which indicate that renovation or restoration work on the building sets the activity in motion. This is commonly misinterpreted as restless spirits not approving of changes to the house they loved. I think there is probably another explanation.

In one case, an old building in Hyde, Cheshire, nothing odd was seen for years, until a wall was restructured, then apparitions were witnessed. When the shop next door was damaged in a fire and repaired, they started again. When the owners finally decided to move, and had to do a few tidying up jobs with hammers and plaster, months of inactivity were broken with the same repeated images of a figure in a Victorian frock coat crossing the room. This cannot all be a coincidence as too many cases feature the same effect, usually without the witnesses ever having read about such matters.

Quartz crystals are known to vibrate and emit an electrical signal when stress is applied to them. They are used today in watches, calculators and lighters, utilizing what is termed the piezo-electric effect. Very possibly, the hammering and other building work on the stone set the crystals into motion and may stimulate a dormant energy field. Other factors may eventually be discovered which also facilitate the same thing. The passage of heavy frontal pressure systems across the atmosphere and excess ionization generated by the sun are both possibilities that have been given some attention by researchers such as neurophysiologist Dr Michael Persinger, from Laurentian State University in Canada.

There must be more going on than we yet know. Probably someone able to detect the signal carried by this energy field must also be present. When all conditions are fulfilled, the latent emotional energy of some past event, held in place by the fields created within the building, are then set in motion by some trigger event. This is then translated into visual form by the mind of a suitably receptive person who chances to be in the room. They are in the right place, at the right moment, to activate, or access, the replay. This may take the form of a hallucination projected onto the room, in effect a video replay of this long-gone moment held in a time-loop.

I suspect that most examples of hauntings (i.e. buildings in which different people report the same phantom doing the same repetitive things) and quite a few one-off apparitions (where a person simply sees a figure in a room but has no interaction or communication with it) probably result from this sort of time-loop.

It is not only visual events that may suffer this way. Auditory ones may be recorded as well.

Researcher Joan Forman was approached by a woman from Norfolk who in May 1973 had visited the Clava stone circle site in the Highlands of Scotland, built by Neolithic people 3000 years ago. She rested against one of the stones to relax, drink in the quiet and entered a sort of daydream or reverie (in other words she altered her natural state of consciousness). Now she heard muffled sounds from the circle site of people shouting and struggling, then saw a group of figures hauling a huge monolithic stone across the ground. The figures had long dark hair and wore shaggy tunics made of matted animal pelts. Suddenly a noise intruded from the twentieth century, as a group of tourists arrived in the field. This caused the time bridge to be withdrawn. The vision and sounds all disappeared.

Again we must wonder whether the enormous emotional energy that must have congregated around the creation and ceremonial usage of this site had imprinted itself upon the fabric of time.

Even when the events are of no great import in their own right these recordings often involve considerable emotional concentration. I spoke to several people at a very old building by the East Gate in Chester. It was being used as a workshop. On more than one occasion heavy footsteps echoing down stone stairs and along a corridor were unmistakably heard by the startled work force. They could only cope with staying in this position by labelling the unseen person 'George' and making light of the presence, but the echoes across time were very real. As one of the workers told me: 'You could follow the invisible figure all the way past you. It was all perfectly normal. The only thing that was different was that nothing was there.'

I discovered that this building had stood for centuries and 200 years ago was a cell in which condemned prisoners were kept. One can readily imagine how one such prisoner in a frenzy of emotion may have paced up and down the stone floor. If, on one day, the conditions were somehow right to allow this event to be partially embedded (perhaps by a combination of the crystalline stone and other factors such as the atmospheric conditions) then this energy may have formed a time-loop which is being played again and again today. Just occasionally, this can now be detected by someone from this century who is working in that same physical space.

In the same way several battles have been replayed in sound only. The war-time raid on Dieppe in France was apparently heard by two English women on holiday nearby in August 1951. It had really happened nine years earlier but their account of the events seemed to match the known explosions, air flight and gun fire in some detail. They heard it as of coming from many miles away, almost echoing around the world. An impressive investigation was mounted by the Society for Psychical Research. Yet nobody else seems to have heard this replay on that summer night, as if it was only audible from the room in which these two women were then located – possibly a time-loop recording trapped in narrow space, some two miles from the centre of the action, and somehow activated by a combination of circumstances.

Mrs Nina Stansfield* lives in a wonderful stone cottage several centuries old on the moors above Todmorden in West Yorkshire. This area is rich in quartz-bearing rock and most of the local houses are built from such materials, probably explaining why it is one of the most active areas for paranormal activity anywhere in the world. More UFO close encounters have occurred within a five-mile radius of this small town than anywhere else in Europe. Whilst Nina has had UFO-related experiences, more immediately relevant are those related to what seem to be extraordinary time-loop recordings. They may offer important clues about what is going on.

Her problems with electricity are intriguing. At times there seem to have been huge power drains creating massive bills and the electricity board were unable to establish why these occur. Equipment, such as light bulbs, tended to overload and explode suddenly after short periods. Sometimes, when using the washing machine, it appeared to suck energy from Nina herself leaving her emotionally distraught. There may be an exchange of energies here to fuel the video and audio time-loop replays which dog the building.

Nina Stansfield has seen people from many centuries inside the house. They appear to be just images that stay only a few seconds then fade. She has even seen animals in the same manner. She also hears many snatches of conversation that just replay out of the blue. She can hear enough of them to know they date from times past, sometimes as far back as the last century when the house was reconstructed. Only rarely, e.g. in a conversation about a mill from 50 years ago, can the words be heard properly. These are so commonplace that Nina has long since learnt to accept them.

Interestingly, her dog also seems to be aware of these sounds. For instance, more than once she has heard the door slam and the sound of her son's motorcycle driving away from this isolated location. The area is so quiet that no other ambient sounds could possibly be mistaken for it. But her son last lived there and made these noises many years before. When her dog hears them he pricks up his ears, looks vaguely around, then returns to sleep. As we can attest, he tends to react violently to any real strangers or noises they make, so evidently he has also become used to these replays. As Nina phrased it to Roy Sandbach and me when we interviewed her: 'I am just like a satellite dish tuning into the past.'

The oddest thing of all that she recalls is when she got a new vacuum cleaner. This became impossible to use because it kept replaying sounds! These appeared to be snatches of conversations from people in the factory who had made it, the driver who delivered it, the shopkeeper who sold it, etc. When she moved it to a different room in the four-storey stone structure, it stopped doing this. Nina had discovered by trial and error that these time-loop replays seem to be very precisely located in space, just like an audible echo might be heard only if you stand in exactly the right spot within a room for the sound waves to be focused correctly together.

Possibly by studying cases such as these we will discover those extra trigger factors. We could compile data about the days when these replays occur. What linked 4 August 1951 (at Puys near Dieppe), 29 May 1973 (at Calva in Scotland) and all the other dates and locations when these various video and audio replays took place? With the aid of a computer survey examining possible factors, such as barometric pressure, atmospheric ionization, etc., we might discover the magical missing ingredient. Then we may be able to 'stake out' potential locations of a time-loop recording and place receptive people there at what we will be able to predict are the most appropriate moments – i.e. when these combination of factors becomes optimum.

If we do this what will result? If these are hallucinations, can only those 'in tune' see them? Or can something real and tangible be recorded for everyone to view? As we will discover, there are excellent reasons to hope that we may be able to build something extraordinary from out of this phenomenon. Indeed, that a Time radio, TV set and VCR may all prove possible!

5

A Slip in Time

EVERYTHING we have talked about so far is, in a sense, an artificial type of time travel back into the past. It is what we might call a trick of the mind. The sights, sounds, smells and sensations may seem genuine, but there is probably no physical transportation through the ages in the way that H.G. Wells envisaged when he invented his time machine. However, there are cases which suggest that literal time travel may not be impossible and it is towards this remarkable evidence that we should switch our attention.

On the simplest level is one example of time researcher Joan Forman's many cases. It is not as dramatic as some of those she has discovered, nor indeed some of those that I have come upon myself. But it makes some basic and telling points about the phenomenon known as the time-slip.

The incident occurred in 1920 when a woman called Mrs Hand, from São Paolo in Brazil was just a young child. She went to visit her grandmother in a hillside cottage on the moors just outside Bury, Lancashire. Playing in the garden she ran inside and found that the building had changed. It was suddenly darker, as if there was less window space, and all the furniture was different and older in design. A door to the kitchen was no longer there. Naturally assuming she had entered the wrong house by mistake, the girl left. As she exited the lighting changed. So she went back into the house and it returned to the strange, old-world formation she had just witnessed. She left to go outside a second time and then, ten minutes later, when she returned for a third peek indoors everything was back to its normal appearance. The light levels were as expected and an aunt was cooking in the kitchen as anticipated.

In effect, Mrs Hand seems to have seen the building in two time zones side by side. But the act of entering the door had worked like a portal or gate through the centuries. Of course, there was no direct interaction here, so it is possible that this was just a time-loop replay like those from the last chapter. If it was presented very vividly to the girl's mind then her hallucination may have become all-embracing and she could feel a part of the scene. If so, it would not be unlike other time-loop replays where the witness becomes completely absorbed by what they are seeing. Normally they do not assume they have slipped through time, merely that they have glimpsed a scene from the past that has intruded into the present. This may be only a matter of degree.

However, there are cases where an interaction does take place and one is left with little choice but to suppose outright deception or a genuine trip into the past. Probably the best known is that of two women school teachers who wrote a booklet, entitled *An Adventure*, to describe their turn-of-the-century visit to France. Here, in the gardens at Versailles, they seem to have slid back in time and walked amongst characters from the era of Marie Antoinette more than a hundred years before. Of course, they at first assumed they had stumbled upon a costume party, until they learnt otherwise. They also witnessed features of the landscape long since vanished but which matched those which had really been in that garden at the correct time in history. That case has been much investigated and debated and I will not repeat its details here. We now know that it is far from unique and that is probably more significant in itself than any attempt we might make to gauge the authenticity of a hundred-year old story.

John Peters* from Durham told me of his remarkable experience. It was October 1967 and he was driving north through the border country towards Kelso in Scotland when he got lost. The road petered out into a track and led through some sparse woodland. Then, thankfully, he came upon a village which, in truth, was little more than a hamlet. He parked the car and went on a brief stroll to ask directions, but saw few signs of life and no other cars. But there was a woman in an odd-looking long grey dress with a shawl draped around her shoulders. She told him where the main road was located.

Mr Peters returned to his car and backtracked into the moors and did find the road that was described to him. He never gave another thought to these events, merely assuming that this was a quiet, sleepy place where some of the residents had not yet fully caught up with the modern age.

On arrival at his destination he apologized for being late and explained why this was. He got funny looks in return, then stark admonitions that he had to be mistaken. John pulled out the map and showed the road that he had taken, thinking it a short cut, then pointed out the position of the village, saying that 'it was so small it is not even named on this map'. His business friends grinned, said it would not be, and told him that it was a good joke. Then they changed the subject of the conversation.

The next day John Peters returned home and drove on the main road until he reached the turn off. Then he decided on the spur of the moment to prove his friends wrong and take some pictures of the sleepy little village. He drove on – and on – across the moors, and past a ruined building or two. But there was no village. He checked the map again. There was no turn off that he could have mistaken or led him astray. This was the only route that could be taken. It was definitely the one along which he had driven yesterday. But there was no hamlet with its mysteriously quiet surroundings.

By now quite baffled, he phoned one of his friends on reaching home and asked why they thought he had been joking the day before. He did not say that he had returned to the road and discovered that there was no village there. His friend explained that there had once been a village along that road, but when a local valley was dammed to make a reservoir about 15 years earlier it and several other very small hamlets were abandoned and destroyed. Some of them were now under water. All trace of others had been more or less obliterated. There had been some ill feeling about it.

'You could have passed through that village if you had driven along the road 20 years or so ago,' he was told. 'But not yesterday.'

In a case such as this, one is left floundering. I must admit that when I first heard it I thought it was a joke just as John Peters' friends had done. Then, afterwards, I assumed it was a tall story told to impress me. The idea of the legendary town of Brigadoon that only reappears within our time frame once every century, came firmly to my mind. However, what I did not then know was that time slips such as this are rare, but hardly unique.

Bo Orsjo from Portland, Oregon, reported one. In June 1974 he had just arrived on the west coast of America from his native Sweden and was unfamiliar with his new surroundings. But he decided to go for a hiking expedition to Mount Lowe, which rises as the backdrop to the town of Pasadena, California.

The day was very misty and he was relieved to reach a green-coloured hotel midway up the slopes in what looked an unlikely spot. He was intrigued by the idea of such an inaccessibly located residence and, whilst it looked more or less unused, there was a maid engaged in some cleaning duties so it seemed to be operating. However, Bo had brought all his refreshments with him, hardly expecting to come across any habitation way up the mountainside, and so he never entered the building itself. He rested and returned to civilization.

A while later he saw a book which described how a millionaire called Lowe had built a railway up into the mountains now named after him, offering tourists a spectacular view. He had hoped to reach the summits of Mount Wilson, but ran out of investment capital and was forced to stop half way, but still erected a splendid alpine-style village, including the hotel. A photograph was in the book. It was the place Bo Osjo had stumbled upon.

There was one problem. Storms, heavy rain, fires and vandals had over half a century ravaged the remains of this splendid village that had, in fact, been abandoned in the 1930s. Reputedly nothing was left but a few scattered bits of overgrown rubble. Osjo could not believe this. He persuaded a friend who insisted that this story from the book was true to accompany him up the mountain. They set off to climb Mount Lowe and find the hotel. The two men reached the location and there was no hotel. Instead there was a camp ground and, exactly as predicted, some cement and rubble out of which many years growth of trees and vegetation now extruded. Clearly nothing had stood there for decades.

Indeed, as the climber was later able to confirm, the building depicted in the book that he saw after his expedition – the one he is adamant that he somehow visited in June 1974 – was burnt to the ground in a major fire in 1937 – 37 years before he rested there after his exhausting climb. Again, if this is an accurately reported event, one is left with little choice but to assume that a time-slip was involved.

Oddly, hotels and restaurants seem to be a magnet for this kind of experience. They are so commonly reported that one is tempted to suspect a kind of modern folk tale that is acquiring a life of its own.

In 1993 Tony Clark first described to *Fortean Times* his own mysterious experience which had occurred when he was a civil engineer in Iran in 1956. He had been to Manjil in the north-west of the country, overseeing the development of a cement works. On his return towards Tehran, the long, dusty road was difficult to traverse and there was little prospect of food despite the day-long hunger of both himself and an Iranian companion. Then they came upon a village, easily recognized by its unique landmark of a pile of stones atop one another. Here was a small roadside restaurant which they entered with delight and not a little relief.

The occupants introduced themselves, gave their names and proceeded to serve up what Tony Clark described as a 'fantastic meal' with several courses. They ate their fill and then asked for the price. This was quoted as being extremely cheap and they paid up and departed. The owner and his wife saw them to their car and asked them to call again and tell their friends. This Mr Clark gladly did, but received much scepticism that such a facility could exist on a little-used road with limited opportunity for trade. However, when three months later he had to return to Manjil with one of his scoffing English friends he naturally carefully measured the mileage and looked out for the pile of stones so they might stop for their repast.

They found the stones and the village at the right location. But there was no restaurant. Despite extensive checking there was nothing indicating that there had ever been one there. Indeed, a villager whom Tony Clark asked told them he had lived there 40 years and in that time there had never been any kind of establishment matching that which they claim to have visited.

If this was a slip through time into a cafe that had once existed at this remote spot, then it begs many questions which often recur in these cases. For instance, why was there no problem paying for the meal with money that, presumably, would have been futuristic? Why did the owner not marvel at a car which, again one assumes, would have been as amazing as a spaceship to the man and his wife? We might argue that this was really a sidestep into a parallel reality. But the nagging questions tend to worry at us.

By far the most astonishing case that I have come across, and for which I have had the fortune to interview two of the four witnesses, poses similar questions and more. What the explanation is I leave up to you to judge, but Geoff and Pauline Simpson from Kent swear that it really took place and acknowledge the problems that I mooted to them with a philosophical shrug. As Geoff Simpson said to me: 'You tell us what the answer is. We only know what happened.'

In late summer 1979 Len and Cynthia Gisby invited the Simpsons to journey with them on a motoring tour of Europe. They would drive south through France and into Spain, spend two weeks there, then travel north back to England.

On 3 October as the sun set they drove on the autoroute just north of Montélimar heading for the border. As darkness was approaching and they were tired they decided to stop at a motel by the roadside. Len Gisby was met by a bellboy here, wearing an odd plum-coloured uniform. The man explained that, sadly, the hotel was full, but if they took a detour down a little track to the south of the autoroute they would find another, small hotel.

It being too late to argue, Len trudged back to the car and they found the track easily enough. It was marked by wall posters advertising a circus. Ten minutes from the main highway, after passing an old police building, they came upon a ranch-style, two-storey construction that looked rather old-fashioned, but welcome enough. The sign outside said 'hotel', so they knew this was their destination.

The occupants spoke no English and neither the Gisbys nor the Simpsons spoke good French. Nevertheless, they made themselves understood and got two rooms. They also decided to eat and ordered about the only thing they recognized from the menu (oeuf – being egg – which arrived complete with steak and French fried potatoes). The meals were served on heavy metal plates in a room without linen or other accepted amenities. Fellow guests were some roughly clothed men drinking beer from large tankards.

After their meal, and with a long journey ahead the next day, the four retired to bed. The rooms were amazing. There was nothing modern about any part of the hotel, just heavy wooden trappings, shutters instead of glass on the windows, a bathroom with plumbing out of the stone age (including soap attached to a metal bar) and no pillows on the beds or locks on the door. In any case, they slept soundly, although Pauline wedged a chair up against the door as she could not break the habit of sleeping without some security.

Next morning they had breakfast, which was simple, comprising bread, jam and disgustingly strong black coffee. There were some curiosities, notably a woman who entered dressed as if she had just come from a ball, with a long sweeping gown. But it was 7 am! Also two police officers, at least they assumed them to be such and from the station just down the road, who were present in the hotel lobby. They wore deep blue uniforms and capes and peaked hats which resembled nothing they saw worn by other gendarmes throughout the several days vacation time in France.

Whilst such things were noticed as peculiar, they were dismissed as quaint customs in this rural backwater – or else, that there was a fancy-dress party, perhaps connected with the circus signs they had seen the night before. Yet when Len and Geoff tried to get the officers to understand that they wanted directions to the quickest way to Avignon and the Spanish border they started to describe a trail that seemed to take them a very long way around. The gendarmes failed to comprehend the term autoroute (French for motorway) and the two couples finally gave up and assumed it must be their own poor accent that was causing the confusion. Gesturing to pay the bill Len approached the manager who scribbled a note saying 19 francs. This was so ridiculous (about £2) that Len Gisby shook his head, saying, 'No – for all four of us'. The manager smiled, as did the policeman to whom the two couples showed the paper. They paid and left without further ado. Geoff remembers whispering to his friend, 'Come on, let's get out of here before he changes his mind.'

They found their way back to the autoroute without any problem and journeyed to Spain, thankful for the good fortune of discovering this hotel. Two weeks later, they returned northwards and, not surprisingly, had already planned to stay overnight at the same place. At those prices who wouldn't! This time the weather was very different – not the hot stickiness and electrical tingling of two weeks before, but dark and very wet. Even so they found the turn off and the circus signs without trouble. But there was no hotel. They drove up and down the muddy route three times but there was absolutely no trace of this wonderful boon to the tourist industry.

Cynthia was quite upset by this, insisting that it 'had' to be there. One of the men quipped that it had probably gone into liquidation and at the prices charged this seemed more than likely. But they soon realized that all trace of it could never be obliterated in just a few days. They visited the motel. Not only had nobody on reception ever heard of such a hotel nearby and just down this turn off, but they denied that any of their staff wore plum-coloured uniforms as described by Len.

By now a little baffled, but simply assuming they had just gone the wrong way somehow or other, the four holidaymakers drove on and stopped in Lyon. Here they paid twelve times as much as they had been charged in their missing hotel; although a still extremely reasonable £25 or so for two rooms and four meals.

The first time the real strangeness of this experience hit them was when they had their holiday snaps developed on return to Kent. Geoff had a 20 exposure film processed by a local store. Len had taken his pictures on a 36 exposure film which had to be returned to the manufacturers. Both men had taken pictures at the hotel, Geoff taking one of his wife in the room by the wooden shutters and Len two outside, his camera having an automatic wind-on mechanism allowing the second to be taken in quick succession.

Neither film was spoiled. But equally neither had any pictures of the hotel on it. I questioned them quite carefully about this, because it is highly unusual to have any form of physical evidence. There was no gap in the sequence of the negatives. The pictures at the hotel were (the men thought) taken half-way into each film. All those before and after came out, correctly numbered as they should be. It was just as if the three photographs that all four visitors insist were taken at the hotel simply were not taken at all.

A research team working for sceptical science writer Arthur C. Clarke was actually able to examine Len Gisby's camera and negatives. They concluded that there was some evidence that the camera had tried to wind on in the middle of his film but had temporarily jammed. At least the sprocket holes on the film at this point were damaged as if they had become caught up in transit. No such evidence was available from Geoff Simpson's film. In other words, this suggests that as they attempted to take the relevant photographs the witnesses believed they had done so, unaware that the film had falled to wind on properly.

The story become public knowledge by accident and the Simpsons said they would have preferred to forget about it. Whilst they have never been interested in making money from this affair, they have had the chance to return to the area and seek out the hotel. Despite extensive efforts and enquiries with the tourist board in Lyon no trace has ever been found. But from a picture in a book the police uniforms have since been recognized as those worn immediately prior to 1905.

What does one say of such a case? Time-slips where a person enters a past environment for a few minutes and then returns to normality are one thing. Cases where four people spend a night in another time zone are quite a different matter. They obviously provide us with a lot more things to think about and, it has to be said, a lot more to swallow if we are to accept that they occurred. I immediately (and cynically) wondered if there was some possible significance in the mention of the town 'Avignon' (phonetically similar to 'having you on'). Was someone perhaps having all of us on? Clearly, this was not a hallucination. It was either time travel in absolute reality or a hoax. No other likely solutions present themselves.

One must again hesitate over why the four people's clothing from the future did not attract attention in the same way as the attire of these people from the past seemingly did. How could a hotel manager accept without question money that dated far into his future? Why didn't a modern touring vehicle from 1979 interest anybody despite (seemingly) being parked outside overnight? (In this era, presumably neither autoroutes nor cars would have existed in the district.)

Questions like these will nag at us. But I have to say that I found the Simpsons to be charming, no-nonsense people and annoyed at the way some had misrepresented their story. One source alleged that when the photographs were processed they showed the group standing in front of thin air where the hotel should have been! I found no evidence of deception, and unresolved questions are not enough to issue charges of dishonesty.

Possibly most intriguing here is that, for what I think is the first occasion, a victim of a time-slip was hypnotically regressed to the incident. The session was with Geoff Simpson (his wife could not be hypnotized) and arranged by Harry Harris at the surgery of Dr Albert Kellar in Manchester. I was present during the experiment.

Harris was a researcher into UFO abduction stories and had the intriguing (but I think incorrect) idea that this episode might be a 'cover story' imposed by the mind. That the four may instead have met aliens that night and the imagery of a phantom hotel was imposed onto their minds to prevent memory of this hidden truth. There was some precedent of 'screen memories' in other cases (e.g. driving down roads that could not be found later). Under hypnosis these had revealed a hidden account of alien abduction (or a fantasy of one – the jury of psychologists remains out on those options). So this was not quite the wild idea that it might seem.

In any case, no hint of any alien component appeared from the hotel story. Geoff Simpson relived the events pretty much as he consciously remembered them, adding little to what we did not know but experiencing the excitement of his night in a hotel that seems not to exist.

Cases where physical evidence to support time travel is provided are extremely rare. But Joan Forman did have another example which has similarities with the case of the night in this phantom hotel. However, that case had an important difference too. Some of the hard evidence survived!

Briefly, Mr Squirrel, a coin collector from Norfolk went to a shop that he had never visited before in the seaside town of Great Yarmouth. He wished to buy some plastic bags to keep his coins fresh, something many collectors do. He described crossing a cobbled street, entering the old furnished store, where all traffic noise suddenly disappeared. He was served by a young woman in a long dress. She had the plastic envelopes but said they were normally sold to fishermen to keep their hooks and she was unaware of their use to store coins. Still she charged him 5p, astonishingly cheap, calling it 'a shilling' (which was 5p in pre-decimal money). This event occurred in 1973, only two years after the currency switch, and many people still used the old terminology. As such Mr Squirrel was not too surprised. Interestingly, however, the shop girl did stare in astonishment at the post-1971 coin he offered up but took it from him without complaint. The bag into which she placed his purchases fell to bits and he threw it away within a few days. The envelopes went brown and discoloured very rapidly, but remained usable. So he kept some.

However, when he returned seven days later to buy more envelopes from the store things were different. There were no cobbles. The interior was far more modern. The woman serving, who said she had no young woman working there and never had done, claimed they did not sell the envelopes in question.

Joan Forman hedges her bets on this case, noting that it seems too inconsistent to be an outright fabrication. She also spoke with someone who claimed to see Mr Squirrel on the day of the second visit to the store, when he first told the story and, she testified, he 'looked as white as a sheet'. But the researcher was also rather worried by the physical evidence.

The actual retention of physical objects across the years is unheard of in time-slip data. The brown, time-worn envelopes that were offered up as proof also had their own problems. Joan Forman sent one to be examined by its manufacturers. They confirmed it as cellulose film, which turns brown and grows brittle over time. They estimated the sample was 10–15 years old. If it dated from the period when the shop looked as Mr Squirrel described it, then the actual passage of time was more like 50–55 years. Of course, one cannot make inferences about the rate of passage of time during such experiences! The company stated that the process to manufacture the cellulose was available just prior to World War 1 but its products were not really found in shops until the 1920s. So there was an approximate match in time; although, importantly, Mr Squirrel did acknowledge that his grandfather had lived locally around the dates in question and was also a coin collector.

All of these experiences, intriguing as they are, have a vague sense of unease about them. It is hard to pinpoint what this is. Perhaps it is just our disquiet about the ability to physically travel through time. However, I must say that the anomalies which are thrown up bother me the most.

Of course, in a field where we do not know the rules, nothing is impossible. Indeed, if Mr Squirrel's story is correct then it provides an intriguing thought. If what he reports can be taken at face value and these little cellulose bags were transposed half a century into the future, then, presumably, his coin went in the reverse direction. Should there not be some little paragraph buried away in the local Yarmouth press or an anecdote told by local people about a shop keeper who was given a strange-looking silver coin with the words 'Five Pence' on it, a date years into the future and a picture of an unborn Queen adorning its face?

6

Sideways Through Time

WE HAVE EXPLORED various ways in which time travel into the past looks possible – through dreams, visions, memories of another life and those rare, but extraordinary, trips that literally take people from today into yesterday. Although there was consistency and much evidence that the past is not irretrievable, especially to our minds, something was not quite so convincing when it came to real flesh and blood jaunts through history.

Perhaps it is merely our inbred difficulty to imagine how such trips can occur without the dreaded paradox. If we ask ourselves awkward questions and get awkward answers, we may fear that the consequence is impossibility. For instance, what if someone mechanically minded had seen the Gisby's car from 1979 as it sat on that quiet French lane in a time before any motor vehicles existed? Could they have gone away, designed a prototype from what they had just seen and changed the future? Or would time close itself into a loop? Maybe that is how cars really were invented in the first place! But history suggests it did not occur like that and, should this happen, then our past would alter. So, would we then all wake up one morning to find ourselves driving Renaults without realizing why? Or is there some mechanism built into the nature of time to prevent this sort of thing?

Our minds reel in the face of these images and the easiest course is to run away, denying that such things could possibly have happened. However, there is another group of events that is different. Evidence is much tighter and suggestive of reality. In these cases people also actually do travel, but seemingly in space. We would call them teleportations – the instantaneous movement from here to there by way of a sort of timeless, spaceless eternity. Nevertheless, if they happen, they are indeed temporal journeys, every bit as much as they are trips from place to place. They must, in effect, be voyages sideways through time.

A legendary case in paranormal research circles is the teleportation of a Spanish soldier on 25 October 1593. He is said to have appeared suddenly in the main square of Mexico City, bewildered and without any memory of how he had arrived, but swearing that he had been serving with his unit on the Philippine island of Manila earlier that same day. This is 14,500 km (9000 miles) from Mexico City, and the fastest possible travel time for the sixteenth century would be at least several weeks distant. The 'proof' of this voyager's incredible step sideways through space and time was supposedly his claimed knowledge of the death of the Manila governor-general. Nobody believed such a wild tale at first. But more than a month later, when news finally reached Mexico City by ship and courier, it was discovered that the governor's death was indeed true and had occurred on exactly the same day that this soldier had been teleported across the Pacific Ocean. If the story was true, then that was the only explanation.

According to ufologist Gordon Creighton, writing in 1965, this was 'no fabrication' and 'there are reliable records of this episode'. However, Robert Rickard and *Fortean Times* decided to check these records out and found something odd. They began well after the incident supposedly occurred. At the time of the death of the governor, there was no chronicling of this extraordinary story. Only a century later had it begun to filter slowly into historical accounts. Even then there is some sign that it was confabulated as time went by, growing stranger with each retelling. In the end, by relying upon seemingly trustworthy accounts from hundreds of years ago, what looks like a truly puzzling teleportation is eventually registered.

Of course, it may have been precisely that. Perhaps the religious authorities decreed that this affair be censored out of history at the time that it happened, because of its perceived associations with witchcraft. On the other hand, it may never have happened at all. True or not, this does show the danger of overreliance upon uncheckable testimony from many years ago. At that time legend constantly rubs shoulders with reality and fact and fiction were mingled daily to create stories of fantastic events. Rather like today's tabloid press.

Fortunately, the evidence for teleportation rests on rather better foundations than one such dubious tale. You will notice some remarkably recurrent features in these cases suggestive of a strange reality.

Case One: Argentina, 1959

In 1959 (date unknown) a businessman was driving south from Buenos Aires, Argentina, and stopped overnight at a hotel in Bahia Blanca. He set off early in his new car but then it was surrounded by a white mass of cloud that descended out of nowhere. He lost consciousness and found himself without any time gap on a country lane but not in his car, which was nowhere to be seen. A truck driver stopped and it was from him that the bewildered man discovered the astonishing news that this was a place called Salta, over 800 miles from his hotel! What was more the time was only a few minutes later than when he had driven away to continue his journey. He had gone through a hole in space/time. The matter was reported to the Salta police, who contacted Bahia Blanca and soon found that the businessman's vehicle was exactly where he last remembered it to be, just a short distance down the road from his hotel. It was empty and the engine was still running.

Case Two: Japan, 1963

Just after 8 am on the morning of 19 November 1963 a bank manager from Tokyo was driving on the Fujishiro bypass north of the city. Two other men were with him in the car. After passing through Kanamachi they had another car several hundred feet ahead of them which was memorable for its colour and make. There was an old man reading a newspaper in the back seat, as all three witnesses attest. They also describe how a white gas or vapour suddenly appeared from out of nowhere and surrounded the other car. They saw it persist for about five seconds and when it disappeared the vehicle had completely vanished.

Case Three: Argentina, 1968

Probably the most remarkable case involved a lawyer from Buenos Aires, Dr Gerardo Vidal and his wife, Raffo. It occurred in May 1968 when they were leaving a family get-together at Chascomus and were heading south towards Maipu. With them were another couple in a second car. This drove a short distance ahead and reached Maipu without difficulty. But the Vidals never arrived. Despite a search back on the road they had simply vanished.

74

Two days later one of the Vidals' friends in Maipu got a call from an Argentine consulate asking if they knew the Vidals. Dr Vidal himself then spoke to assure them that he and his wife were fine but utterly mystified. It was alleged that they had been driving south of Chascomus when a dense fog appeared on the road ahead. The next thing they knew it was daylight and the car was halted in a country lane. They had no idea where they were but felt as if they had slept for a long time.

Asking for help and directions as they drove they soon discovered the astonishing truth that they were in Mexico more than 6,400 km (4000 miles) from where they had started. Forty-eight hours had passed, but it is probably impossible for them to have driven that distance in such a short time even if it were feasible for us to envisage a gigantic hoax.

On the night of their disappearance a man was admitted to hospital at Maipu stating that as he drove along this route a strange fog had fallen onto the road surrounding him but remained only a short time, leaving him feeling unwell. The watches of all those in the cloud had stopped suddenly.

Case Four: Argentina, 1968

There were several other slight reports which have an element of doubt associated and which may be considered copy-cat hoaxes. There has never been enough data to check them out and so their status remains confused.

However, one case is interesting. An 11-year-old girl called Graciela was on the doorstep of her house at Cordoba on 4 August 1968. It was mid-afternoon. Nobody was out playing so she decided to go indoors to watch TV. But then suddenly a white misty cloud appeared on the path ahead of her. It surrounded her and she lost all sensation until coming to in a square full of people. She wandered around lost for some time until darkness fell and she knocked on the door of a house pleading for help. The occupants took her to the police station, where it was found that she needed hospital treatment for shock and she was able to tell her fantastic story.

If what Graciela says is the truth then she was moved some miles across the city in an instant. There seems to have been no time lag. We might imagine that she wandered off and cooked up the teleportation story to explain her misbehaviour. But could she have walked so far in just a couple of hours and, even so, would she describe the same pattern as other cases?

Case Five: Essex, 1974

The Avis family of John, Sue and their three young children were returning from a trip to relatives on an October night. They were rushing home through the deserted countryside on the eastern borders of London to catch a play on BBC television. Suddenly a thick bank of green fog came out of nowhere and straddled the road directly ahead. It was too close to avoid running into this and as they did so the radio in the car began to spark. Fearing an electrical fault John, who was driving, yanked out the wires. Moments later there was a bump, as if the car had leapt over a hump-backed bridge and crashed back to the ground. The mist was gone. They were disorientated, as they seemed to be further down the road and almost home.

Upon arrival John stayed outside for a few moments to fix the wires on the radio and told Sue to go indoors and switch on the TV to prepare for the coming drama. However, she soon told him the shocking news, which they confirmed with others over the telephone. The TV drama had ended. Indeed, the TV channel had gone off air for the night. It was – impossibly – almost two hours later than it should have been.

Because of subsequent events, involving the intervention of (as it happens very objective) ufologists and the use of regression hypnosis, this case is widely regarded as an alien abduction, or spacenapping. However, the facts as just presented are the initial conscious testimony of the witnesses and reflect the essence of the time and space anomaly which confronted them. Any attempts to unravel these events tend to impose an interpretation which may, or equally may not, be valid, but is speculation.

You may well be concerned as to why so many cases happened in Argentina. In fact, as you will see, there are cases from all over the world and in recent times, but I suspect that part of the reason for the concentration may be that this region south of Buenos Aires is an area where space-time effects somehow manifest. There is evidence from elsewhere that there are other pockets of the phenomenon. A small area of Cheshire countryside between Daresbury, Preston Brook and Helsby has generated at least half a dozen known examples of cars and motorcycles being teleported between 10 and 40 miles and up to six hours disappearing from recall.

As another example, in September 1973 a 21-year-old Italian man living in Bedford, England, had been to a dance in Northampton but on his way home a strange thing happened. It was about 2 am and he was at a village called Little Houghton. He recalls seeing a glowing mass through the windscreen directly ahead of the car, then nothing. The next memory he had was of wandering about in a daze, as if he had been in a car wreck. He was at Bromham Bridge, more than 30 km (20 miles) away, and it was now 7 am. He was not inside his car, but staggering along the road-ways soaking wet, even though it was not raining at the time. Eventually his car was located. It was in the middle of a field near Olney. The vehicle was locked and, despite the heavily mud-strewn field, there was no indication of how it had got there. Not a trace of a tyre track was found. It could not be driven from the field and had to be towed out by tractor.

This story is interesting enough, but ten years later, in February 1983, the village of Little Houghton was at the centre of another mystery. I am virtually certain that the witness on that occasion, whose name was Peter Rainbow, had no knowledge of the previous case which had achieved no media attention at the time.

Peter had been riding his motorcycle to visit his mother in the village when just before 7 pm he lost all power to his engine and lights. Thinking a fuse had blown he got off the machine and endeavoured to fix it with a piece of silver foil (an old biker's trick). But this did not work so he was about to give up and replace the spent fuse when he spotted a silvery mass beside the road, swaying from side to side. All at once there was a blink in reality and the mass was gone. Peter noticed the fuse was missing (he later found that it had not been placed into the motorcycle). However, he was holding the ignition key, which previously had been in the lock. He put this in and turned it and the engine fired normally. Driving on from the (unknown) countryside spot where he was now located he reached Little Houghton moments later to find that 1½ hours had somehow passed.

These British cases were interpreted by many in a UFO context, i.e. assumed to be 'missing time' in which the witness was abducted by aliens. Much interest in teleportation has purely been to seek this option. By applying regression hypnosis and taking the witness to the event a 'memory' of what 'actually' occurred (i.e. a flight on board a spaceship) is then sought.

Such a technique was applied in none of the first four cases we discussed, and in truth there is very little reason for us to do so. The link with UFOs and abductions is a tenuous one, quite often more the product of investigator urgency than the result of anything more concrete. However, the temptation is inevitable. A case with remarkable similarity to the first from Little Houghton was reported from eastern Europe in 1992 and was also subject to the, probably misleading, UFO evaluation.

The location was Szekszard in Hungary, near the shores of Lake Balaton. Late one evening, probably in January 1992, a woman was driving her Trabant car in this area when she observed a strange glow ahead of her. At the same moment she lost consciousness and recovered to find the car inside a field with the engine and lights off and now failing to operate.

The woman struggled from the car and dragged herself painfully towards a light in the distance, which proved to be the guard house at the perimeter of an industrial complex. They took her to hospital where it was discovered that she was suffering from shock and had some blood on her foot from numerous unexplained small lacerations.

The police were called and discovered the car in the middle of the field. But the problems only really began with this discovery. For the field was a mass of snow and had been like that for several days. Yet there were no car tyre tracks leading into it from the nearest road, itself some distance away. What was more the Trabant's door handle was melted and almost welded shut, although there were no other signs of a fire.

The comparisons between the Olney and Szekszard cases are obvious. In both instances the vehicle somehow moved into the middle of a field without leaving any evidence of normal traction. How did it do so?

It is not only human beings and cars that experience these odd transportations. There have been reports of animals, particularly cattle, being teleported in mysterious circumstances.

In January 1984 a very distinctive cross-bred bull turned up overnight inside a completely locked barn at Chillerton on the Isle of Wight. The farmer is adamant it was not there the night before.

Similar incidents have been reported elsewhere, e.g. at Clifton, Kansas, a few months later, where a distinctive animal appeared from nowhere.

An incident which combines many of these elements also helps to illustrate why the UFO theory is so widely contemplated. Indeed this particular incident is often regarded as one of the world's most significant alien abduction stories. But let us review the facts, and add a little known sequel to it, and we might be forced to reconsider.

The events befell a West Yorkshire police officer, Alan Godfrey, at just after 5 am on the morning of 28 November 1980. He was about to go off patrol but decided before doing so to take one last look for some cows that had been causing a nuisance to a housing estate in the small town of Todmorden. Nobody knew how they had got loose from their field, but loose they were and he wanted to find them.

As he drove down Burnley Road Alan spotted a rotating mass on the road ahead. It spanned the width between the hedges and lamp standards, causing the bushes to vibrate in the wind generated by its motion. He stopped the car to stare at this amazing sight, then there was the by now familiar blink in reality. One moment he was there, in the car, engine and lights on, staring at the object, the next he was mystified to find the engine and lights off and that he was still in the car but it was an unknown distance further down the deserted roadway than he had been just an instant before.

Subsequent reconstruction by ufologists from Manchester suggested that the teleportation from one position to another might not have been quite so instantaneous. There was a possibility of a ten-minute period of time missing from the officer's recall, although that is far from certain as he was not paying undue attention to the clock under the circumstances. There was some sign of a struggle (his police boot was split as if he had been dragged along the ground). All in all it was a bizarre experience.

In a case such as this, particularly in this day and age, it is quite common to interpret the hovering gaseous mass as a UFO, to assume that there really is missing time and to use hypnosis to explore the gap further. Alan Godfrey was subjected to hypnosis six months later and he did then describe seeing aliens inside a strange room. But he was commendably honest about the status of this memory. He told me that he had read about UFOs during the intervening months and that, whilst he knows what he consciously saw on the road that night, there is no way he can judge the reality of what he said of later events under hypnosis. It is very possible that this was a fantasy.

If it were, then the interpretation of this case as a classic alien abduction becomes rather dubious, of course. I personally suspect that to be true of many others like it. 'Missing time' is, I fear, often more apparent than real and what is actually missing is memory of a teleportation. Our mind urges us to presume that such a journey must have taken a finite period and, as few of us would know if 10 or 20 minutes were unaccounted for when we are so entranced by such an experience as this, then it becomes easy to talk ourselves into assuming that there was missing time when there may not have been any time lapse at all. In other words, what is quite clear is that a strange force was encountered that was apparently able to move the car from one location to another, possibly without any passage of time.

Yet the most intriguing aspect of the Todmorden story is rarely mentioned when this case is discussed as a UFO abduction. The cows were found as daylight broke, in a wet muddy field, well away from where they were last reported – and inside an area that was completely locked! There had been heavy rain earlier in the night and the field was very muddy. Had the cows entered by any normal route they would have left some evidence. There was absolutely no sign of how these cattle got into that field.

Just like the puzzling effects on Alan Godfrey's police patrol car, the mysterious transportation of the cattle suggests that very strange forces were at work in that quartz-rock-infested mill town on that peculiar night.

There are many more cases that I could cite which seem to indicate that teleportations like these are occurring on a regular basis. Here are just a few examples.

On 8 October 1981 an American couple on tour with a widow of a REME Colonel were driving on a deserted road across moors in the Salen Forest on the Isle of Mull. They slowed to a halt so that one of the Americans could take photographs and were suddenly surrounded by a mist that came from nowhere and began to thicken and solidify around the metal body of the car.

Dwight, the would-be photographer, had faced bandits in Turkey, but was still scared. The car began to vibrate and the one witness saw how the mist blackened and moved about them in an odd way, like sentient shadows. Then it just vanished. They got out from the car to find themselves lost on a quiet road and disorientated. The boot was open and contents thrown about.

They were still on the island and found the main town Tobermory easily enough. However, the sun had moved well across the sky and they had apparently lost several hours of time during this unknown journey. Both the Americans had quartz-powered watches and these had stopped working and had to be reset by a jeweller. The quartz clock in the car had also stopped.

On 15 October 1983 Catherine Burk from Bellwood, Pennsylvania, USA was driving her car on the highway at about 9 pm. Suddenly she heard a whirling noise and observed a silvery mass with a protruding hemispherical base directly above her car. As it came within feet of the roof the vehicle was sucked up off the road and carried like this for some distance. Mrs Burk was trapped against the door as the motor vehicle tilted sideways. Then the steering, engine and lights all failed. The hold was finally released and the car crashed onto the deserted road with a dull thud.

Regaining her senses Mrs Burk found that the car engine would still not operate, although some minutes later it was successfully revived. She suffered considerable trauma, including headaches, affected vision and itching blisters. She also had to wear a neck-brace for a time. She did attempt to secure payment of her medical bills but her insurance company decreed that this was an act of God – but could not say exactly what kind of act it was!

An interestingly similar case created world headlines on 20 January 1988 when it struck a family travelling on the Eyre Highway in Western Australia. Like the incident at Bellwood we have evidence of a strange floating mass sucking a car from the road and creating damage extremely like that which hit the vehicle on the Isle of Mull. But this Australian case occurred at 4 am on a long, flat stretch of road when a time distortion may have simply not been noticed, had one taken place. Knowing exactly where and when you are is notoriously difficult on the longest straight road-way on earth.

The Knowles family were on a long drive east from Perth to a family reunion. They spotted a yellowish white mass ahead of them snaking about in a misty, sinewy path. At its closest it was directly on top of them and made a humming noise. The car shook violently and there was a weird distortion of the voices of the occupants as they spoke in increasing panic – their pitch seemed to be altered like a Doppler effect (which changes the sound of a speed-ing car as it approaches and then moves away from you).

At one stage the occupants of the car on the Eyre Highway insist it was pulled into the air by the floating mass. Then the car crashed back to earth and slewed off the road, one tyre shredded by the impact. The family of four leapt out and hid in bushes fearing that the thing would still be there. The contents of the boot were later found strewn about the roadside.

Eventually, having changed the tyre in great haste, they hurried on to the next town – which was some miles down the monotonous, straight track at Mundrabilla. Here they reported their experience to others, who later brought in the police. A fine yellow dust was said to have been deposited inside the car and a smell like electric arcing or bakelite was also in evidence.

Several other drivers on the road that night reported how they were suddenly buffeted off their course by strong blasts of wind that hit them out of nowhere.

Only two weeks later, on 9 February 1988, a case occurred which achieved no publicity at all and was reported directly to me. The witness was a truck driver who at the time was looking for farm work and was walking along a quiet road near Oswestry in Shropshire. It was 8 am on a clear, sunny morning. He saw an elderly woman with a dog that she had let off its lead and which was running across a roadside field barking and yapping as if at an unseen intruder.

Puzzled the man looked again and saw now that a peculiar cloud was straddling the gap between hedge and kerb. It was yellow-white, like a sickly mist and gave off a noise like a cross between a motor vibrating and the wind blowing fiercely. The entire thing was rotating violently. Several other factors accompanied this weird sight. There was an uncanny stillness and silence, a powerful odour of musky gas and a prickly sensation like all the fine hairs on one's body standing on end.

However, the most frightening thing of all was that the dog (a spaniel), oblivious to any danger, ran into the mysterious cloud and vanished. Its woman owner was screaming hysterically and the truck driver was desperate to calm her, without being sure what to do. He looked at the rotating cloud. It was at least 9 metres (30 feet) in diameter and was remaining in one place. But of the dog there was no trace. It had simply been swallowed up by this peculiar fog.

However, as suddenly as it had appeared, the mist began to dissipate, then vanished altogether. It went like smoke disappearing. In the place where it had been was the dog, flat out on the ground with its body on the footpath and head on the kerb. From a distance it looked dead.

Rushing over to the animal several things were noticed. It was panting heavily and its eyes were bloodshot. It was also soaking wet, although the morning was completely dry. Indeed, steam was rising from it as if it had just stepped out of a sauna. The man carried it to the woman's car and put it on the passenger seat, covered by a blanket.

A few days later a check revealed that the dog had recovered, but it died not long afterwards from what appeared to be natural causes, which may, of course, have been accelerated by the severe trauma that the animal obviously went through.

In the cases discussed in this chapter there have been a number of consistent features. The most common of these is the sudden onset of the glowing mist or fog that appears from nowhere, condenses around a vehicle (or in the last case, an animal) and then rapidly disappears. This seems to be intimately related to whatever mechanism is responsible for taking the car or animal on a trip through space, time and, perhaps, both.

As we have noticed the degree to which this occurs does vary. Often the displacement is small enough to be hardly noticeable, indeed sometimes very probably not noticed. I strongly suspect that undiscovered encounters with this mysterious mist will exist when no time or space dislocation was recognized by its victims.

In other cases the space or time dislocation is understandably, if I suspect incorrectly, assumed to be the product of a UFO kidnap. There are also cases on record from the past where similar 'kidnapping fogs' were attributed to the works of fairies and demons.

Of course, in those extreme cases, such as the Vidals apparent transportation across an entire continent and thousands of miles, the effects are all too obvious and inexplicable. But there are other clues that may prove just as important. For example, the effects on watches – but only, it would seem, on those with electrical or crystal mechanisms that may be affected by electrical or magnetic fields.

The appearance of water condensing onto the bodies of those who are transported is also found in several examples. And then there are the smells and tingling sensations which are suggestive of electromagnetic radiation. Indeed, some of the physical effects with which witnesses later suffer – for example, itching rashes, bloodshot eyes and headaches – are also symptomatic of exposure to mild radiating energy fields.

There seems little doubt that some naturally occurring force that manifests as a glowing cloud is appearing on a regular basis in the world around us. When someone gets too close the consequences may be amazing. They seem to include the possibility of a trip through space or time.

As I write the most recent case that I have come across was investigated by Ken Phillips and Judith Jaafar on 8 August 1992. A family drove towards Milton Keynes in Buckinghamshire hoping to do some shopping on a bright Saturday morning. At Hockliffe on the Bedfordshire border all sounds just vanished and a sudden thick mist and very localized rain shower hit their car. They were – 'in the blink of an eye' as one witness described it – 13 km (8 miles) from where they had been – 'we just seemed to be there'.

For the next hour or so the husband and wife in particular noticed strange sensations. Their bodily coordination was shot to pieces. In trying to open the car door the wife missed the handle and the husband could not operate a petrol pump. There was also a tingling sensation and effects on their vision. The two young children were oddly subdued as well.

So severe were these effects that later during the morning they actually drove to see a family member and asked what looks like a very weird question – 'Are we really here?' They did so, because it had crossed their minds that the strange things that were happening to them all (which thankfully soon faded) might be best explained if they had been in a car crash and were now dead!

They were not dead, of course. But they may have had an equally fantastic journey through time and space.

7

The Oz
Factor

WE SEEM TO HAVE CONCLUDED that 'real' travel through
space and time may rarely occur as the result of some, probably
natural, energy field with which we may chance to come into contact if we are unlucky. Its radiating fields appear to affect our perception of time, space and reality to such an extent that we may
even mistakenly conclude that we have 'died'. However, just as
often during our look at these strange time-related experiences we
have seen how the mind plays the fundamental role. Indeed, we
have found that an altered state of consciousness seems to be
required if we are to reach backwards, sideways or forwards in time.

There is something which recurs in these cases and which
appears to be the key to this altered state. I call it the Oz factor, after
that magical fairy-tale land where reality is different from our own.
The onset of the Oz factor is the clue that points the way towards
this place that we are seeking, not a land that time forgot but,
instead, a land that forgot time. You cannot miss the Oz factor in
operation. It cries out to you from so many different experiences
where time seems to take a back seat. Here are just a few examples.

Mary Latimer from Illinois, USA, told me of the apparition that
appeared in her bedroom at dawn one morning. It turned out to be
an image of her grandfather, then several thousand miles away in
California. He was waving goodbye at the precise moment when, it
was later revealed, he had suffered a fatal heart attack and died. It
was as if he had made a pit stop on the way to heaven to bypass
space and time and give one last farewell to a favourite relative.

However, as important as this interpretation are the feelings that
Mary expressed from the moments whilst this took place: 'It felt as
if time had disappeared. I was stretched out between now and
eternity. There was only my mind and his mind bonded by love.

The experience could have lasted fifty seconds or fifty years and I would not have known the difference. Somehow in this place that we inhabited there was no such thing as time. That is the lasting memory – the one truth – that stands out from all of this.'

Timelessness is one crucial element of the Oz factor, but another major thing to look for has featured often in the cases that we have seen so far. It is the sudden isolation in space, where all the normal ambient sounds and sensations disappear. Johnny Caesar, an actor and stage musician, who now has a role in the TV soap opera 'Emmerdale', told me of the night that he was electrocuted during a performance at Aviemore in Scotland. The audience thought it was part of the act when he crashed to the ground. In fact, he had been catapulted out of his normal everyday state of consciousness and into the Oz factor where he went through what is commonly called a 'near-death experience'. He saw his body from a vantage point above and the frantic efforts to save his life.

But Johnny describes what it felt like to float free of time and space: 'There was no hurting involved. It was a lovely, warm sensation. I saw them working on me like it was a TV screen and I was watching a play. There was no sense of sound or panic. I was not hearing, more sensing what was happening. But I was quite alert and was just not taking it in. I felt that it would be so easy just to drift away.'

Eventually, to all our good fortunes, a sense of purpose and desire to return reasserted itself and Johnny Caesar fought back. He made a full recovery, but in this out-of-body state he, like thousands of others, came to see that no words can adequately describe the isolation from space and time that one is suddenly immersed within. It seems to be like taking a bath in forever.

We would not expect to discover this effect in UFO sightings, but it is there in almost all close encounters that involve more than the sighting of a light in the distant sky. Retired headmistress Eileen Arnold well illustrates this point. Her encounter was in spring 1944, three years before the term flying saucer was first invented. So what she saw that afternoon was simply a mysterious vision in the sky, quite unfettered by today's preconceptions about aliens and spacecraft which, researchers feel, heavily colour what people describe.

The mind, after all, is not a camera, but a living, active, thinking process that changes dynamically and evolves. We see not only what is there but what is filtered through our own perceptions, beliefs, expectations and past experiences. Often what we 'see' may not be at all what is there.

Eileen was walking down the High Street in Cheltenham, Gloucestershire, returning from a prenatal check-up. The pavements were busy, the road alive with traffic as the sun shone just after noon. Interestingly, Eileen says that she was in a 'particularly sensitive state' that day, possibly because of her thoughts of the impending birth. She was doing many things 'on impulse' and having what felt like empathetic links with other people, even strangers she walked past. Then, something made her look up and she saw a spectacular object sailing across the sky. It was unlike any UFO you might have heard about before, being akin to a giant porcupine ejecting quills into the air as it glided through the atmosphere.

Of course, this phenomenon could not drift across a busy town in the middle of the day without hundreds, perhaps thousands, of people seeing it. But in fact it seems that nobody else saw it. This isolationism is one of the most intriguing, yet most common, aspects of the UFO close encounter.

Of the period whilst she watched this thing move silently above Cheltenham, Eileen records: 'There was a time lapse and an environment lapse – which I don't have a name for. Time slowed. The extremely busy road altered. Traffic and people had completely vanished but I did not even see the road and pavement . . . all I 'saw' apart from the UFO was the rooftop over which it appeared.'

The Oz factor in UFO cases often manifests like this. Sometimes the witnesses talk about sounds (like singing birds or traffic noises) fading away. At other times they are suddenly alone on a road normally full of other cars or a pavement that should be full of people.

I suspect that this tells us something important about what is taking place. Eileen Arnold's words convey the impression of her consciousness tuning out all the normal sensory input and superimposing this one vision that was entering her mind right on top of the world around her. It is rather like a ghost that may be a hallucination, but, if so, it is one that gets overlaid on top of the bedroom in which it appears and dominates the scene. Here the spectacular UFO was somehow detected within Eileen's mind and was so overpowering as an image that it blocked everything else from her awareness and effectively made the sounds and sights disappear.

This moment is sometimes called by poets who have experienced it – the 'eternal now'. In that period Eileen melded with the experience and everything else, including space and time, was secondary in importance.

Eric, from Essex, who spent 28 years in the police force and so, as he stressed, had passed rigid medicals and was not suffering from any sort of mental illness, described to me how it feels to have the Oz factor take over. This was in one of several visions, of the past or future, that he has experienced: 'The atmosphere would abruptly change, the surroundings becoming remote. Activities around me slowed down and I felt apart from the immediate environment. When I say that the surroundings became remote I do not mean that they receded as happens in a fainting fit. I felt it was me who was leaving the prevailing conditions.'

A drama teacher from Israel told me of her frequent visionary experiences, such as one in which she was catapulted spontaneously through time whilst reading an article about Stone-Age excavations: 'Suddenly I was there. It grew darker and I felt myself floating or falling through space and then was rooted in this reality. It was not a dream. I was standing on wet earth. I felt this and I could see everything. But my actual surroundings from the modern day had disappeared. I did not know where or when I was. Time had lost meaning. But I had travelled somewhere else within my mind. I was there.'

The Oz factor can surface in many such unexpected places. Fred Rayser reported his experience in the summer of 1937 whilst taking part in a sprint race at Spring Valley, New York. Being an accomplished runner he was given a handicap, which in practice meant that he started up to half a lap behind other sprinters. In a one and a half mile event this was an almost impossible gap to close. But he tried, urged on by his family.

Fred reports how, as he rounded the final bend, things changed. His intense focusing on the one aim seems to have altered his state of consciousness and pushed him through a strange barrier. He says: 'The light appeared to change from bright sunshine to a muted glow. It was as if I had entered a translucent tunnel.' He noticed the other runners ahead had all slowed down and 'seemed to be moving in slow motion'. He weaved through them all with ease and felt himself fly past, bursting through the tape with so much adrenalin pumping inside him that he could have done the race again. Indeed he had set a record time, leaving the rest flailing, making up an unprecedented amount of space by appearing to bypass time.

The stress of a race does not compare with that of a battle. Yet

Herbert Lehmann of Euclid, Ohio, relates his experience in October 1944 when he was part of a troop of US marines trying to capture a small rocky bluff before the Japanese could do so on the Palau Islands in the Pacific Ocean.

He had crawled his way to the top of the ridge before a team of Japanese soldiers made it from the other side. Both they and his own troops were firing machine guns and lobbing grenades across his prone body as they battled for supremacy. The victors would survive. The losers would all die.

Suddenly, Herbert reports: 'All sound faded until everything sounded as if it were coming from far away. It was as if my head were under water . . . I also seemed to be floating . . . I remember that I was not frightened, a condition that was remarkable considering all the noise and confusion around me.' Not surprisingly, he had made the same mistake as did the family who had the teleportation at Hockliffe 48 years later and assumed that he had been shot and was now dead.

But Herbert Lehmann was very much alive. He was, in a sense, shielded by the Oz factor, cocooned by these strange sensations that demarcate the shift from one state of consciousness to a timeless other. Seconds, minutes, years later the effect vanished and he was back on the ridge with the battle subsiding. They had won. The Japanese were vanquished. He would survive.

As for time-slips pure and simple, the Oz factor is found there too. Joan Forman talked with C.H. d'Alessio who was walking the streets of London one evening in 1975 when he seems to have been projected into the future. He saw strange cars floating past on cushions of air and roadways with a synthetic, silvery feel to them. He was utterly convinced that for a few moments he touched base with the next century and experienced this part of the city as it would then be. During the episode he felt all the traffic sounds disappear, everything became muted, time lost its hold on him and he was in the same oddly calm and trance-like state that Herbert Lehmann had felt.

Indeed, if you think this through, you may have already concluded that it may be possible that Eileen Arnold's fantastic porcupine drifting over Cheltenham was a vision of some future aerial technology of our own. We must expect that if time-slips backward are feasible and precognition of tomorrow's trivia by way of dreams can occur, that literal slips through the fabric of time into some future environment might also be a possibility.

My term Oz factor is only one that is convenient. Many

psychologists and philosophers have noticed this same effect from their own studies of human experience and tried to define words of their own. Read the works of Jung, Freud, indeed almost anyone who has looked at the fringes of human adventure, and you will find that they have come across it. Phrases like 'the peak experience' or 'cosmic consciousness' have in the past been applied where the timelessness and spacelessness of this ability is detected.

However, it is difficult not to seem mystical when speaking like this. What I wanted to do was to anchor this state of mind very firmly in real experience. I have surveyed what witnesses say in this wide range of paranormal phenomena, both obviously time-related (such as time-slips) and not so obviously related (such as near-death or UFO close-encounter visions). The pattern is the same and you can readily build up this portrait of the state of consciousness which seems to facilitate their occurrence.

That is the first step. Giving it a neutral name to capture its essence of magical transportation was the next move forward, hence the Oz factor.

We must move on from here. It helps a little to say that the Oz factor is a set of symptoms denoting an altered state of consciousness in which the normal bonds of time are freed and the mind senses the universe as it really is and can wander through those corridors of forever.

However, it is very probable that there is a physiological change within the brain which corresponds with the Oz factor. We know that various emotional states are related to the flow of certain hormones that can affect our neural network. Secretion of adrenalin is obviously important in stress situations and can boost physical strength, for example.

We do not know exactly what parts of the brain or what chemical agencies correspond with emotions such as love, but recall that we found very early on that emotion was a key to all strange experiences. Without some sort of emotive link time and space barriers will not be defeated and no relevant information will gate crash into your consciousness past the doorman that we have set up to protect our waking selves from the intrusion.

As we probe deeper into the workings of the brain, thanks to new techniques such as the CAT scan, allowing us to map brain activity in a person who is asleep, awake, in a coma or doing just about anything, then we can piece our way towards a recognition of what is involved. By that process we may eventually isolate the chemicals and the parts of the brains that are stimulated and that correlate with the Oz factor state.

There are already some promising signs from work by psychologist Dr Serena Roney-Dougal who has been assessing the actions of the pineal gland, long thought by mystics to be a source of the hidden 'third eye' and which, they suggested, allows spiritual insight. A chemical trigger, via naturally secreted hormones such as serotonin, which is already receiving some attention, may provide a way forward.

For example, in several well-studied cases where animals have been known to detect an earth tremor some time before it happens (they enter a sort of Oz factor state and go eerily quiet in the hour or two beforehand) it has been speculated that serotonin emissions in their brains are stimulated by very small levels of some radiating energy that emerges from the changes within the earth as the tremor builds up towards its peak.

It may be possible to imagine a time when this process can be understood and controlled and an early warning system applied to suitable humans to serve as living, breathing earthquake detectors. Then it will look as if we have harnessed precognition. Of course, that will not really be what is happening. But then, if the Oz factor is similarly a product of chemical reactions, the inter-relationship between what is physical and what is psychic, may ultimately prove to be irrelevant.

What we need to do to make time travel practical is conduct physiological, psychological and parapsychological studies to find what it is that causes Oz to come about. If we can isolate these things, then we will be well on the way towards producing a device that can stimulate time travel.

This will probably be something that can induce the Oz factor more or less to order and give our mind the gentle push that it may need to set sail across the oceans of time and space. When that is possible, and it will be in the near future, then a whole new generation of explorers will be set free. Unlike Columbus or Magellan they will not be confined to the boundaries of earth – all of space and time will be at their disposal.

8

Dreaming of Tomorrow

'I RARELY DREAM (or at least rarely remember dreaming), but at the time in question, possibly because I was rather on edge as a result of family stress, I was having more frequent and vivid dreams than usual,' so began Michael Senior from Warwickshire, in his letter to me.

Michael's memorable dream from 1974 was almost stunning in its triviality. He was eating a meal at a seaside restaurant with a view over a distinctive harbour. He did not recognize the place, but the colours and sensations of the imagery swept over him and he took in the pattern on the tablecloth and the scene as a bald-headed stranger walked in carrying a bright red book. The man took the seat next to his own. Then he pulled out a newspaper and started to read.

Michael awoke feeling that there was something more to this mundane yet powerful vision, but was quite unable to grasp what that something was. In the end he let the dream memory fade, but had worked it around enough in his mind that some of it etched its way inside and did not simply disappear. Ninety-nine per cent of all our dreams will vanish within twenty-four hours of our having them unless we make some conscious attempt at storing them in our memory banks.

Six months passed and the dream was relegated to Michael's subconscious. It was obviously one of those things that mean nothing at all. Then Michael went on holiday to a part of Scotland that he had never visited before and suddenly, there he was in the restaurant, overlooking the harbour, sat at the same table and remembering the dream. It did not come back like a vague sense of *déjà vu*. The dream memory hit him 'with crashing vividness'. He found himself speculating whether the bald man would enter and, sure enough, moments later, the stranger did just that and the exact scene from Michael Senior's dream many weeks before was now re-enacted before his eyes.

Michael stared at the man, hoping that he too might have dreamt of that meeting and show signs of recognition. But there was no hint of that. The moment passed by and slipped into history. This was just one more irrelevant incident such as shapes all of our lives. Nothing significant seems to have been related to what in other circumstances we would have to call a non-event. The two men never even spoke to one another.

But, think about it. Was it really as insignificant as it seems? In the dream Michael Senior felt there was something important but could not grasp what that something was. When the event occurred, he went through the same emotions. Indeed, as we can now see the precognition was of that moment when Michael would grasp this stunning irrelevance of time. He actually sensed ahead of itself his own inability to comprehend the fallability of everything we are taught about the orderly structure of the universe. For if moments of trivia can leap six months ahead, what about powerful dramas? Is there anything fixed within the fabric of time?

I suspect this was actually quite an important incident to have seen ahead of time after all. But what of this next case?

Michele from Derbyshire described what happened to her. She dreamt that she was making her tea, which mostly comprised banana sandwiches. Then the phone rang and it was her boyfriend's best mate who explained that her boyfriend had just had an accident, had fallen off his motorcycle and was now in hospital. Michele's response was not, as might be expected, concern, depression, or to ask many questions as to how he was. It was the wholly inappropriate remark 'Oh yes!' and to burst into laughter.

That was the end of the dream.

Michele recalled it well enough to tell both her boyfriend and his mate when they met the next day. They smiled about it. However, a few hours later, after Michele had returned home, she was indeed making banana sandwiches when the phone rang and – sure enough – it was her boyfriend's best pal speaking the exact words from her dream. Without thinking, Michele said, 'Oh yes!' and began to laugh, convinced that they were both playing a practical joke on her. But they were not. Her boyfriend's best friend had forgotten the dream. What he was telling her was the truth.

As it turned out the accident was not serious, but as Michele was quick to spot the dream itself introduces a more serious problem. When we considered journeys into the past we had to contemplate changing history by some act that might be carried out in the past. Of course, to anyone in that past time our present would be their future, so this changing of what to us would be history would become an alteration in their future!

If you have not been foxed by all of this, then think more about Michele's dream. She foresaw a puzzling reaction that she exhibited to a real event – the phone call about the accident. Yet that puzzling reaction directly came about as a result of her dream of the puzzling reaction! We find ourselves running round in circles here, seeking the answer to the question science considers immutable: which was cause and which was effect? Did the dream create the future response to the dream, or did the future response exist already and ripple back through time to generate the dream? Hinging on what seems like such a minor dream and fairly innocent question is possibly one of the most important implications. Is our future mapped out for us in minute detail, or can we somehow skirt around it and alter what is yet to be?

Most physicists would be terrified by the implications of Michele's dream, since it seems to challenge the very essence of cause preceding effect and the one-directional flow of time. We seem to move from the past through the present and into the future. What is past has happened and so is utterly unchangeable; what is going to happen is yet to be and so quite unpredictable. Somewhere in between is an indecipherable moment – the now – which clearly must be a finite period and yet can never be measured. As soon as we grab hold of it to put it alongside our ruler or to measure it with our atomic clocks it has gone and been replaced by the next now.

In fact yesterday and tomorrow seem to be bridged by a span which has zero length – which, literally interpreted, would tell us that there is no span at all, and that 'now' is just an invention. Yesterday IS tomorrow!

Oddly enough, experience from the real time-travel stories we have met seems not to contradict that apparently absurd idea. Maybe the past and the future are one. When we are free of this inhibiting invention of ours that there is a 'now', then we experience the timelessness of reality in what I call the Oz factor. As a result we experience the illusion of time travel.

Of course, the irony is that time travel is not happening at all by this definition, because if past and future are one and the same we are merely exploring different aspects of a single, cohesive whole that is already in place – rather like going to Africa or Australia without needing to travel back in time in order to do so. Indeed, it may prove to be even more ironic than that.

Georgina Mills from Lancashire told me of her terrifying dream. She was in her own living room talking to her mother about the death of her father. When she went downstairs she was so relieved to discover that her father was still alive that she hugged him, saying nothing. As she told me: 'You cannot exactly go up to someone and say that you have just dreamed that they have died.'

A few weeks later Mr Mills did die, suddenly and tragically, and Georgina found herself experiencing the dream within that living room. There were broad similarities, but the scene was not an exact duplication of what she had previously foreseen. There were several key differences, e.g. not the same people were present in the room with her.

I think we can see why this was. Probably all that slipped past the doorman from the future was the raw emotion of that scene in the living room – the tragic impact itself. The imagery in Georgina's dream was largely constructed from her mental store of information, clothing that terrible memo from tomorrow in vivid pictorial form. To do this she would have used images that were in her mental store already, i.e., in effect, memories of the room itself and the people expected to be in it.

The weird implication of this is that, whilst in one sense Georgina Mills did have a vision of the future, she did so by way of images from her past. Only the script came from tomorrow. The players within it – as well as the scenery – all came from yesterday. Precognition involves seeing into the past. This idea of premonitions in a dream being a mixture of input from the past and future further illustrates their interchangeability and may only seem odd to us because of the way we think about time. Free of that heavy obligation to see time as a linear progression it may all make perfect sense.

The problems of deciphering what is precognition are well displayed by a dream of my own which I regard as having seen the future, although that is probably debatable.

It was a night in early October 1984 and I was having a very restless sleep. At 04.13, as I recall from the digital clock, I got up to have a drink and returned to bed. In that period before 8 am I had a very vivid dream.

In my dream I was with a party of police officers outside a large building. There was talk of a terrorist siege. Suddenly something was thrown across my path, like a grenade, and voices screamed that it was a bomb. Then it exploded and I saw clouds of dust and debris rise.

Although this was a vivid dream, it was not a particularly lucid one. In other dreams I have sometimes awaken unable to tell, like Georgina Mills describes, whether I had just dreamt something or was remembering a real event. I need to check to reassure that my dream was just a dream. This is a sign of lucidity. I have noticed that the more lucid the dream state the far closer to reality its images are. In a situation like my 'bomb' dream, when merely vivid but not lucid, there is often much distortion to the key theme.

Almost as soon as I got up that day I put on the TV news, which was an unusual act. For the next couple of hours I was saturated in images of the terrible thing that had occurred that night – 400 km (250 miles) away from where I slept. The IRA had planted a bomb at the hotel in Brighton where the Conservative party was having a conference. The death and destruction, whilst thankfully limited, was something that hit the country very hard.

Now, in most circumstances, had this dream been reported to me by somebody remembering it months or years later, two things would probably have happened. As time went by the dreamer would have followed a natural tendency to make the dream more like the experience that it seems to predict. We all do this. I prevented it here just by adequately documenting in writing all of the details the moment its importance was recognized.

The second likely consequence would have been to have assumed that this was a premonition of the disaster. Fortunately, the fact that I am certain the dream occurred between 04.15 and 08.15 that morning proves this to be untrue, for the explosion had already occurred an hour or so before I went back to sleep.

I have often found from my experiments that this is a good way to increase the chances of dreaming across time. If you wake up in the night, fall asleep again, and take careful note of dreams during that second sleep, then they are more likely to contain future-related imagery.

Of course, there was no normal way that I could have known about this disaster when I had my dream. It was still paranormal. But was I sensing emissions through space, rather than time?

I suspect not, because when I analysed the confusing images in my dream it is easy to see that many of them relate to scenes that were presented on the TV news that morning – smoke rising from the façade of the building, the huge police presence keeping

The idea of time travel has captured our imagination. The movie trilogy *Back to the Future* was a huge international success and suggested that an inventor could build a time machine into a De Lorean car.

A Victorian concept of future technology shows how basic trends can be seen, but not in the exact form that TV would later adopt. (*Robida, 1881*)

A city in the late-twentieth century according to popular views of the future in the 1920s. It has a few points that match the reality but applies mostly imaginative guesswork. (*Cuningham/Grey's Cigarettes*)

Modern science fiction speculates freely on the nature of time travel into the future or to change the past.

The Aborigines have a sophisticated view of time. Sacred sites include Ayers Rock in Northern Territory, Australia, which is viewed as a repository of power from the dream time. The idea of energies locked up in rocks is now being taken seriously by Western scientists.

The Aborigines see the dream time as a powerful serpent coiled inside the living essence of the earth, as depicted in their traditional artwork.

Bill Waddington (Percy Sugden in Granada TV's 'Coronation Street') has empathetic experiences which allow his mind to tune in across space and time. (*Granada TV*)

Reuban Stafford's death certificate. But 100 years later Ray Bryant has memories of his life as if they are his own.

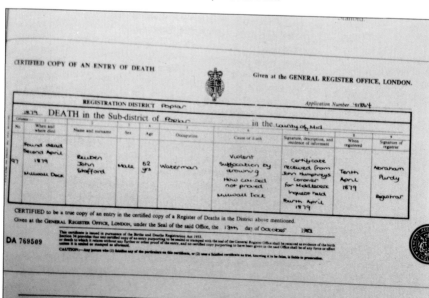

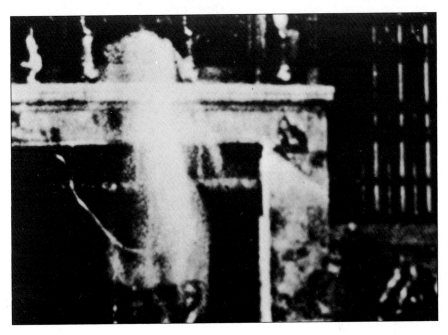

A 'ghost' photographed beside a church altar. Nothing was seen at the time. A trick of the light – or of time?

A controversial photograph of a phantom monk with its feet below the level of the ground, located where the floor level was in this church years before.

Jenny Randles investigates the moorland location above Todmorden in West Yorkshire where mysterious sounds from the past are constantly being replayed like a tape recording. (*Roy Sandbach*)

The green bank of mist that straddled the road as depicted by the family at Aveley, Essex, who encountered it in October 1974. Typical of that seen by many witnesses who, as in this case, find themselves suddenly transported through time and space.

Jenny Randles with a team from London Weekend Television researching the quartz-infested rocks around Todmorden, West Yorkshire, where countless strange experiences and time anomalies have been reported. (*Roy Sandbach*)

The split police boot of Alan Godfrey who lost time and was transported through space in November 1980 following yet another mysterious event in this valley.

The artwork typical of the visual creativity of witnesses who experience frequent 02 factor states and consequent time anomalies. (*Judith Starchild*)

The highway that weaves past the airport at San Francisco, California. Scene of a frightening dream in which the future was seen and altered for the better.

The Niagara Falls on the Canadian/USA border. In a dream a disaster was foreseen involving the *Maid of the Mist* boat seen here ferrying tourists into the maelstrom. Nothing happened. But soon after, freak conditions caused the tourist facilities to be closed. An off-key premonition or a coincidence?

The grassy knoll in Dallas, Texas, scene of the November 1963 assassination of President John F. Kennedy – foreseen by several people in dreams. Psychic Jeanne Dixon successfully predicted it in advance in her syndicated prophecies.

Michel de Nostradame ('Nostradamus') – whose prophecies are interpreted and reinterpreted constantly through the ages.

A volcanic lava plug on Lanzarote. Jenny Randles was conducting a remote viewing experiment from the dormant volcano on shore during the time of the chain reaction involving the Colombian volcanic eruption.

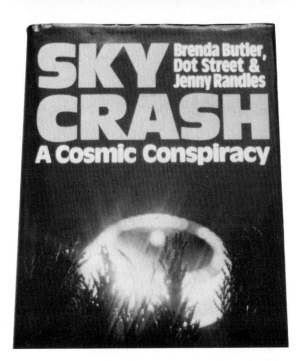

Sky Crash – the predictively titled book in the shuttle chain-reaction story.

The shuttle is prepared for take-off at Cape Canaveral, Florida, after flights were resumed following the *Challenger* catastrophe.

Psychologist Carl Jung who developed the theory of the collective unconscious mind and, with physicist Wolfgang Pauli, the concept of synchronicity – meaningful coincidence.

The USS *Eldridge* supposedly involved in the Philadelphia Experiment that transported it through time and space. (*US Archives*)

Fate magazine reports on Tesla's secret experiment. The electronics genius was also reputedly a mastermind behind the Philadelphia Experiment. (*Fate Magazine*)

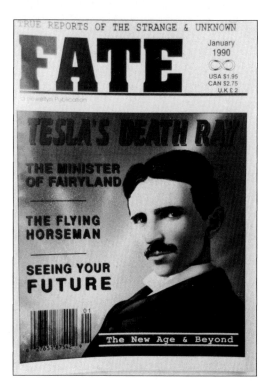

The photograph taken by the Jaraslaw brothers at Mt St Clements, Michigan. They later said it was a hoax, but it depicts exactly the same kind of object seen at several other locations spread across almost a quarter of a century. Were they all the same time-travelling vehicle? (*Jaraslaw & Jaraslaw*)

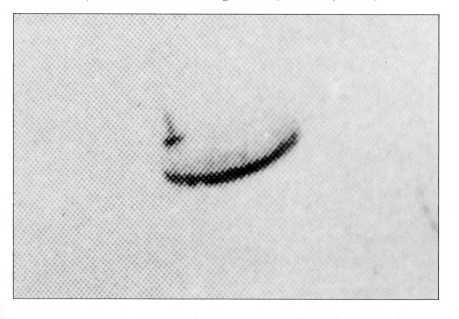

The object photographed by the Trent family at their farm in McMinnville, Oregon, in May 1950. Is this a time machine?

The bizarre triple image on a single frame captured by a scientist crossing the Williamette Pass, Oregon, in 1966. How did this object phase in and out of space and time in this way? Or is it just a moving image of a signpost as sceptics allege?

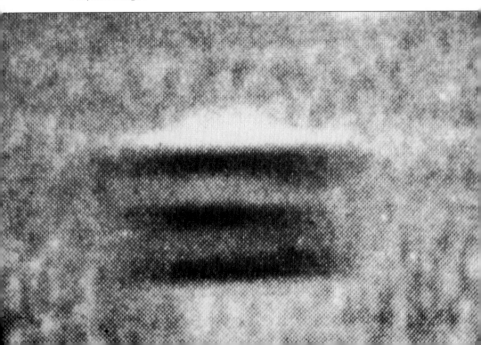

The Colorado River cuts into the Grand Canyon in Arizona, USA, exposing dozens of layers of rock laid down over millions of years of the earth's history. Every layer represents a period of time longer than human beings have walked the earth.

The desire to find methods to time travel is strong. In the final chapter I propose some suggestions that can be attempted.

Crop circles in Wiltshire fields. Unusual anomalies throughout history such as this might suggest the presence of time travellers and can be sought out.

A brief note in a dream diary – which was my first success. In practice, diary notes should be more detailed.

FRIDAY
Mar 8 1968

Peper factory by bridge. Bus stop on

reporters and sightseers at bay, and so on. In fact images from after the explosion and the time when I first saw its effects.

My dream was less of the bombing of Brighton than it was of my immersion within the huge saturation of TV news coverage about its aftermath several hours later. As very often, if not always, seems to be the case, the precognition is of your own state of mind at the time of recognition of the event rather than of the literal event itself. What crosses time is an emotive response – your emotive response – rather than facts and figures.

This is a very important discovery about seeing into the future, because it will later help us to understand why virtually all attempts to prove this ability in the scientific laboratory have been doomed to failure. Being rather difficult to control this feature in any experiment, scientists tend to leave out emotion from their research – when, in truth, this is the one critical factor that is essential to make their experiments succeed!

My Brighton bomb dream also shows another difficulty. How can confusing images such as this be used to predict an event? If we cannot recognize when a dream does foresee the future (although its strong emotional content seems to be a major clue to watch for) and if the dream will match the reality to a greater or lesser extent, but is rarely an exact mirror of it, then is it even possible to use these things as warnings? Indeed, even if it is, can we alter what we foresee? And if we alter what we foresee, then the dream fails to become a precognition – so have we foreseen anything at all?

The question of looking into the future is thrust upon us by experience, not by active design. We are having these dreams presumably because they are a part of our function as a human being. In which case we must plough our way through the confusion that they reflect and work out why this is happening.

Confusion was certainly the way of it with another of my dreams, but it is worth relating since it illustrates a further difficulty to be confronted. When the dream is attempting to portray a future event it is rarely able to do so as a snapshot or video. Instead it uses the language of dreams – which is often highly symbolic.

Psychologists from Sigmund Freud to Ann Faraday have written popular books telling you how to decode the confusing images of a dream. As an example, I once dreamt of being trapped in a flood and escaped by climbing onto a grand piano. I consulted two psychologists about this daft image. One told me that I regarded music as a way to set my consciousness free. Another pointed out that dreams often use puns and suggested that I was afraid of being drowned by emotions and needed a boat to sail away. My mind had word associated with boat and come up with the shipping line – P & O – and from that created a 'piano' for me to escape upon!

I have no idea which, if either, psychologist was correct, but there is plenty of evidence that dreams do use symbology in this bizarre manner, so you can see how awkward it can get.

The precognitive dream occurred on 4 January 1983. Two things of note had happened. The *Daily Express* had written a piece on my work, but I had no idea when it would be published that week. Also the Merseyside radio station, Radio City, had asked me to do a 15-minute slot on the paranormal which could become a series. I was excited about this radio challenge and that night it must have been on my mind.

I had a very curious dream that was unusually well recalled when I awoke. I made a careful record about it. Quite often nothing results from doing this when it seems advisable, but occasionally the scheme pays off. It did so here.

In my dream I was involved in some sort of war between two rebel camps and had deserted to join one side in preference to the other. I felt rather concerned and worried about all this and left the room, clutching a press cutting which for some reason was very important to me. One rebel handed me a golden wedding band saying 'I'll give you a ring', but as I tried to escape a car blocked my path. It had the licence plate MCK . . . P (I could not recall the numbers). In my dream I worked this through, saying to my dream self 'This car is from Preston. It is a real number plate.'

After that I awoke to the trilling of my telephone. The call was from Radio Red Rose which had seen my interview that morning which had appeared in the *Daily Express* and wanted to conduct an interview. I said I would have to check with Radio City, but afterwards started to worry. Then, as I sat down to record the memory of my dream, its significance hit home.

My worry was simple. Radio City (based in Liverpool) over-lapped to some extent with Radio Red Rose (based in Preston but by which I had never been interviewed before). The day before I had pledged allegiance to City, now if I 'defected' to Red Rose because of the press cutting they had seen, then I might be acting like a rebel deserter. Red Rose had actually said to me, as their parting words – 'I will give you a ring' – the exact final line spoken, but symbolized in my dream as a literal wedding ring.

As for the link with Preston, that was the oddest of all. As I had somehow done in my dream state I could now actually work out that MCK . . . P was indeed a real Preston registration mark issued in 'P' year (1975). I had worked in a job for a time in which I needed to know such useless information and had an amazing wealth of knowledge about car licence plates. Preston issued 161 marks in 'P' year and MCK was one of them (it was NOT issued in the preceding or following year, for example).

However, the precognition was not quite through. Later that day, when I did get the go-ahead from Radio City and when Red Rose called from Preston, the man who interviewed me was a well-known sports commentator whom I had no idea worked for that station. His name was Hugh Macklin. MCK is, in fact, the only one of the Preston combinations issued in 'P' year for which all three letters occur in sequence in the name Macklin. It took me ages to work that out – when awake. My dream had done it in seconds, in order to symbolize the name of an interviewer I would not talk to until several hours later.

The mind rather boggles in the face of that sort of complex precognition. It also illustrates just how hard it will be to try to use such dreams to predict (and prevent) future events. We do not know which dreams do contain future images. We do not know how literally they preview the event. They will also probably portray the person's mental state when they first discover the event, rather than be an actual preview of the event itself. They will have at least some degree of distortion introduced by factors such as the image store in the mind of the dreamer, from which the dream is clothed and then acted out, the symbolic nature of most images in the dream state and the distortion introduced by memory because dreams are notoriously difficult to recall and most people do not make a full documented account of them immediately after they happen.

Faced with problems such as these, you might think it barely worthwhile to try to use dreams as a means of precognition. But the rewards are potentially enormous. The desire to know what is going to happen is strong.

The awful dilemma of dreaming the future is well illustrated by Phyllis Hudson of Wisconsin, USA. When her son was born in July 1948 she began to have a series of horrible nightmares that he would drown.

The three dreams were not identical. Whilst always vivid they had a different method of drowning, in a run-off water pool at a zoo, in a well or in a pool. But in each case the boy died.

Not surprisingly, Mrs Hudson kept a very careful watch on her son every time he was anywhere near water. Extreme precautions were taken, e.g. covering a well in a house they moved into. Whenever near a swimming pool he was never left unattended.

Then, in 1961, aged 13, the boy mowed the neighbours' lawn and was left unsupervised. Knowing nothing of the dreams, of course, he suggested when he had finished that he clean their swimming pool. Through a terrible accident where his face mask detached when he was cleaning the bottom levels and he was unable to swim to the surface, he paid for this kindness with his life.

Despite three vivid warnings and the taking of every conceivable precaution in response to their different threats, the future was not altered. You might argue that the boy's death was postponed by the response to the earlier dreams, e.g. covering the well, but if so, this rather implies that fate cannot be cheated for very long.

This story has a frighteningly fatalistic ring to it. However, contrast it with the experience that befell Marilyn Mayer, who in 1968 was a classical musician in San Francisco, California.

One day in February she was due to travel with three others to a recital, but at around 6 am had a vivid dream. In the dream she was in the car with the others, as expected, going through the journey. As they passed signs for the airport the driver screamed 'Oh my God!' and a car slewed across the central divide and hit them suddenly, in a head-on impact. She awoke in panic, relieved it was just a dream. Yet the emotional force was such that she was in little doubt that this was a warning of a real crash that would occur to them in a few hours time.

However, determining there was nothing she could do to escape fate she pressed on and when one of the other musicians arrived Marilyn told him and he grinned. Calling it a 'stupid dream' he suggested they sit the same way as she remembered in the car and bluff it out. They went to meet their lift.

The driver was 20 minutes late in arriving, explaining that the fourth passenger had experienced unusual problems parking in the city and so been late meeting up. However, as they set off all went well, although rain now fell heavily. They reached the sign to the airport and Marilyn exchanged glances with the man she had told of her dream. Suddenly, the driver yelled 'Oh my God!', just as in her dream and they screeched to a halt. A police patrol was flagging them down. Ahead, in the terrible conditions, a car had just slewed across the central divide, hitting the first car in their lane head-on and killing everyone inside. Had they been on time, rather than a few minutes late, it could easily have been them.

When she reached home Mrs Mayer had a phone call from an old friend in North Dakota, over 3200 km (2000 miles) away. At 7.30 that morning she had woken from a dream in which she had been called out of a rehearsal to be told that her friend (i.e. Marilyn) had just been killed in an auto smash on the Bayshore Freeway in California. The friend was making plans to attend the funeral in the dream when it ended. Upon waking she was so upset she tried to warn Marilyn, who had already left for her fateful ride.

As they talked on the phone, the doorbell rang. There stood a second friend. He had driven 30 km (20 miles) to reach Marilyn Mayer because of a terrible feeling of doom that he had awoken with that morning. He just had to reach her to warn her to take care.

What are we to make of this? Obviously, if we take it at face value it suggests that the event of the impending accident rippled ahead through several hours to one of the four people in the car and two of her close friends. Note how in the man's case, he had only detected it as a raw Psi-emotion and had no explicit vision or dream. Also how her friend, thousands of miles away, had foreseen her own personal role in the incident – i.e. the moment she learnt the news. She did not actually foresee the accident. This is exactly what we would expect from our research so far.

Of course, the real question is why an uninvolved person in North Dakota could be touched by this temporal ripple and yet three of the four people in the car had no forewarning at all? Or did they? We cannot know what their dreams might have shown had they been better at recalling them. Equally, we have the intriguing possibility that the musician who caused the delay may have not done so by happenstance. What if the event had been detected at a subconscious level, but this did not consciously register?

We know that our subconscious mind can engineer behaviour patterns – the so-called 'Freudian slip' is a case in point, when we say something unintended but which represents a feeling that our subconscious is expressing. Motivational behaviour manipulation of this sort is equally tenable upon quite straightforward psychological grounds. If deep down the crash had been foreseen but that image could not be forced past the doorman into conscious awareness, it is likely another route might have been found utilizing the raw instinct level of our being. The problem finding a parking spot was the physical outcome of this and it may well have prevented the quartet leaving the city in time to meet with that fateful accident.

There is some evidence for this. Physicist Danah Zohar found cases from the files of the SPR (Society for Psychical Research) in which a dream, or a vague recognition, was not heeded as such but provoked an unusual vigilance that has prevented disaster. In one example, a yachtsman dreamt that he was in danger of being rammed and sunk by another vessel, discovered that fog had descended, but still ignored the dream. Yet a feeling nagged away at him and made him instinctively go aloft just in time to prevent catastrophe.

We speak often of intuition, a 'gut feeling', a 'sense of knowing' and a 'warning voice'. This seems to be a very real and common phenomenon. It is the subconscious working upon data that it has detected ahead of time.

American researcher William Cox is the only one to have put this into practice with a clever experimental design. His work really needs follow up and duplication as it provides exciting evidence of this ability.

What he did was to collect statistics about train loadings and compare the numbers who travelled on a train and in the relevant carriages in the days preceding a fatal accident to that particular vehicle and timetabled journey. He found that in every case available for study there were fewer passengers than would normally be present in the very coaches that were due to be wrecked during the lead-up period to that accident.

The odds were more than 100 to 1 against these figures being due to chance. Whilst not startling, they are certainly more than suggestive that passengers were somehow being subconsciously manipulated away from the doomed carriages as the rippling energies from a forthcoming disaster gathered in strength and were accessible to their consciousness.

An interesting case which shows the value of this subconscious 'instinct' concerns Charlotte, a woman from Tacoma, Washington, USA, who in July 1980 was daydreaming as she did the dishes. Outside her husband was mowing the lawn. Suddenly a strong Psi-emotion got past her doorman. She knew something awful was going to happen. Then a powerful and graphic Psi-vision projected itself across her eyes. She saw masses of blood at a hospital.

Of course, Charlotte's logical reaction would be to think that her husband was about to have an accident and respond accordingly. Interestingly, however, she did not do so, e.g. banging on the window and attracting his attention. Quite unconsciously she stepped back and away from the sink and said nothing at all. Which was just as well. Moments later the mower hit a large stone and threw it into the air. By a sheer fluke this struck the window, shattering it into a hundred deadly pieces. But for stepping backwards Charlotte would have been horribly injured and her premonition of blood and a hospital would have rapidly been vindicated.

Here there can be no doubt that behaviour was manipulated at a subconscious level because of an awareness of a coming event.

Hollywood actress Lindsey Wagner seems to be another perfect example.

On 25 May 1979 she was at Chicago's O'Hare Airport ready to check in to an American Airlines flight when that old devil mood overtook her. Swamped with a horrible feeling without explanation that nearly made her physically ill, she simply had to cancel the reservation for both herself and her mother. Which was a life-saving experience, for the DC-10 lost an engine immediately after take off and crashed into a horrific fireball at the end of the runway. Nobody on board survived.

We must wonder how many passengers had similar gut feelings, or even dreams, but did nothing about them. At least one case is known of a woman who had a dream of disaster and cancelled her ticket.

Nevertheless, the reality of trying to use dream precognition to good effect is also well displayed by this same event. A man called David Booth from Cincinnati, Ohio, actually dreamt about it ten times shortly before it occurred and was undoubtedly fired up enough to try to prevent it from happening. The story of how and why he failed is very important.

Booth's first dream was nine days before the accident. It was unusually specific. He was able to tell that it was an American Airlines plane, from its colour and logo and, whilst not identifying the type, that it had three engines. He knew something was wrong with one engine, that sound was muted and saw exactly how it plunged to earth. He described the scenery too.

He had almost identical repeats of this every night until 25 May. By 22 May he was no longer in any doubt that all of this meant something disastrous, so he called the local office of the Federal Aviation Administration (the FAA). He also talked to the service manager at American Airlines, who recommended he talk to a dream psychologist.

One of the FAA employees filed a report on the conversation on the 22nd. This confirms that Booth did see several key details, including the airline, type of aircraft, the time of day as mid- to late-afternoon, the reason for the crash and description of where it happened. Several false assumptions were also made, notably that the aircraft was landing and that the number 40 was part of the flight. The FAA had also guessed (understandably but wrongly) that the crash was supposed to be local and that, as Boeing 727 jets land at Cincinnati, it was meant to be one of these.

When the accident did take place David Booth was left with a terrible sense of frustration.

His dreams were probably 75 per cent accurate. Some of the images seem to relate to later TV images of film of the crashing jet captured by chance from a tourist at the airport. Yet 279 people had died and the combined resources of Booth's powerful and better than average precognition, the FAA and American Airlines had all failed to do anything about it.

Could they have prevented it? Had they tried they would likely have grounded the wrong aircraft – i.e. all Boeing 727s flying into Cincinnati. Yet, even if the exact airline, airport and aircraft type had been foreseen would it really have been feasible to stop all flights that matched the description for an indefinite period? When scientists use sophisticated equipment to warn of a high level of danger of an earth tremor but nothing happens after a few days there is soon a demand that life not be disrupted for a possibility. If the predictive mechanism is a series of dreams the likelihood of any positive action resulting is minimal.

Yet we still ask ourselves that crucial question – what if? Could David Booth's warnings have saved all those people? If the flight had been prevented from taking off on 25 May 1979 would the crash still have occurred at a later date? Would the future have been altered? Clearly hundreds of people would have gone on to have lives that would have affected millions of others simply because they survived that day. So, in a very real sense, had these dreams been heeded the entire world would now be different as a direct consequence.

Major disasters such as this one are a rich source of precognitive dreams. That is much as we might expect. The powerful ripples from a deeply emotional tragedy which affects countless people is bound to have more chance of affecting people's consciousness ahead of time, because more will be touched by it. The event itself will also be given greater prominence. Whilst it obviously depends upon other factors in addition, the simple reality that a major catastrophe will be screened in front of perhaps billions of people on TV news bulletins is bound to increase the likelihood that some of those will foresee that moment when they come into personal contact with the tragedy via such a media report. As such, all that we have learnt in our studies so far, tells us that these are the instances when precognition will be most marked and that the spread of mass media publicity may well be making precognition more common than ever before.

When I sought other examples of dreams of the Brighton bombing akin to my own I found several instances. Dreams previewing the assassination of President John F. Kennedy in November 1963 are quite common. The murder of Beatle John Lennon in New York in 1980 also affected a lot of people ahead of its occurrence. The SPR has several well-documented cases of people who saw the *Titanic* sink before it really did go down. The story is repeated many times with similar incidents.

However, one of the most studied cases is that of the awful tragedy that struck the Welsh mining village of Aberfan when in October 1966 a coal slurry, weakened by constant rainfall, collapsed and fell onto the local primary school claiming the lives of 144 people, mostly children. It was an event that touched the world like few others.

The precognitions were clear and unhindered by time and space. People all over the world have reported them. Dr John Barker, a London psychiatrist, collected 60 excellent cases and in almost half was able to verify through independent witnesses that the dream was told to others before the disaster. These dreams run the complete range and again highlight all our difficulties. One was highly symbolic (just a team of black horses riding down a hill slope). In another school children wearing Welsh national costume were seen floating up to heaven, a graphic presentation of coming events but in a way few would have understood beforehand.

In such examples, the essence of the event was obviously detected and then cloaked in the symbolic language of dreams to such an extent that we can recognize the significance only after the event

Another incident involved a woman who visualized the rescue operations and later, when watching the TV news, observed the same scene replayed. This again illustrates how the dream was about her own mental link with the tragedy and not with the tragedy itself.

However, the most remarkable of all concerns one of the children. How many of those killed actually had premonitions we cannot know. But one of them, nine-year-old Eryl Mai Jones, had two. A fortnight before the disaster she told her family she was unafraid of death as she would be in heaven with Peter and June. In the mass grave she was buried between them. Then, the morning before the tragedy she told of a strange dream in which she went to school but a black thing had fallen over it and the school had disappeared. Of course, the intriguing thing about this last dream is that, one presumes, Eryl Mai's conscious mind probably never knew what actually happened when the slurry struck and killed her. What aspect of her own future mental state did this image of the missing school and the black sludge thus come from? It almost hints at some post-death conscious awareness of the aftermath of the tragedy.

Building upon these observations it seems theoretically simple to perfect a system whereby dreams can be recorded, centrally registered, somehow filtered and analysed and used to issue warnings in the same way as a combination of seismograph, electrical field and gaseous emission readings from inside the earth can be pooled by scientists to predict maximum periods of threat from an earth tremor.

In the wake of his collection of data on the Aberfan disaster, Dr John Barker decided to do just that and began what was called a Premonitions Bureau. In 1967 it received 500 assorted dreams, all bar a handful of which were incoherent or inconclusive. Few, if any, seem to have been in any way usable to predict an event up front. Of course, nothing as traumatic as the Aberfan affair was to occur that year.

Dr Barker tragically died in 1968, but ITV science journalist Peter Fairley continued the work into 1970, latterly being promoted via *TV Times* magazine. Did this succeed? The answer would have to be no. Around 1000 more dreams were received and some did correlate with actual events. However, the concept of devising a method by which advance warnings could be issued was never perfected. Probably the experiment needed more time to gather strength and it suffered, of course, from the fall off in submissions when the initial burst of publicity waned.

An American counterpart was set up at a dream research unit in Brooklyn, New York. This received thousands of reports in the late sixties and early seventies and probably had a few more spectacular individual successes. However, the fraction of submitted dreams that in any way seemed to relate to a real prediction was probably little more than 1 or 2 per cent. Also, given the computer technology of the day, to use this to devise a system to study patterns and issue warnings would have needed at least ten times as many dreams sent in each week than arrived at the project's zenith.

Some interesting work was conducted in 1974 by a clinical psychologist, Dr J. Orme, based on 148 well-documented cases, including some from Dr Barker's work and the US registry. What Orme wanted to do was plot the frequency of multiple precognitive dreams that seem to be referring to key events (like Aberfan) as a function of the time gap between when the dreams occurred and when the premonition was fulfilled.

From the cases featured in this chapter you will probably be able to suggest the outcome – that most dreams occur close in time to the perceived event – but this research was important because it showed that there was a mathematical relationship. Most dreams were focused around just a few hours before the incident. As time increased to about two weeks ahead of the event they fell away markedly. Only very few occurred further back than 14 days from the foreseen incident. The graph showing dreams against time followed a smooth curve indicating that there was a well-defined and predictable scientific order to the time structure. As an event came closer and closer to fruition, the dreams rose towards a peak.

These mathematical results, which could now be extended and improved upon, offer a very important weapon to any modern-day attempt to run a computer-based premonitions clearing house. Given today's powerful computer technology, the strength of the media that could promote it and our knowledge of patterns such as the above, a new venture is worth attempting. Its greatest hurdle is the need for patience. If backed by any media source it is likely to be expected to produce results immediately. The chances are that this will not be possible. The project would have to run for several years and efforts would need to be made to keep its supply of data coming in. If it were working efficiently at the point when another catastrophe of the emotional magnitude such as Aberfan were to strike, as sadly we must accept as inevitable, then the project should come into its own. Perseverance may pay dividends and we would certainly learn a good deal more about the nature of dreams that are perceiving the future.

I did get the opportunity to run a pilot operation thanks to *She* magazine. But it was very small scale (with only a few hundred thousand readers), ran for only three months, and as such had a limited scope to function properly.

Project Dreamwatch ran between August and October 1988 and submissions arrived under the fairly strict premise that I set. This said that they were to be vivid dreams that the participant felt powerfully enough about to consider previews of future events, but also of events beyond their own personal lives, i.e. of national or international note. Most dreams are of personal matters, as we have clearly seen, and this was as a consequence a severely restrictive imposition. But it was a necessary one for the experiment.

Given these limitations 29 dreams were received and documented immediately, thus provably being ahead of any event they might foresee.

Of these dreams it is safe to say that approximately half were completely wrong. They predicted specific events, e.g. the deaths of named people or cited disasters, at given dates, none of which were fulfilled.

Another category, accounting for all bar seven of the remaining dreams, were inconclusive. They sometimes suggested future events that have not occurred but theoretically still could, or they were vague enough to maybe fit an event that did happen but which cannot fairly be scored as a success. One was an interesting submission of a dream that may have described an event that occurred the morning after the letter was due to be sent but because of a postal strike the recorded posting was after the incident.

The remaining seven are interesting enough to review briefly.

In one an earthquake was described but it was stated to occur in South America. Three days later a major tremor did take place, but in the Himalayas. On balance this has to be termed a miss, even though some details did match.

A natural disaster more accurately foreseen was a catastrophic flood that struck Africa, well-described in a dream about two months beforehand. But is that too long a gap given the frequent occurrence of flooding?

On 16 September a dream warned of danger to North Sea oil operations. The dreamer allegedly had foreseen the tragic Ekofisk rig disaster and tried to warn a worker there. Six days after the dream was logged at the project, the Ocean Odyssey marine disaster occurred, which may be considered to match the warning, although fears about the rigs were general.

Perhaps the most specific dream was of a boat containing school children which sank with heavy loss of life. This was logged on 23 August, six weeks before a real accident with many similarities did occur.

An intriguing dream was of a fairground ride disaster at a fun park. The dream was logged on 27 August. Two days later a BBC film exposé of safety dangers at funfair rides was aired and, being a subject newly raised for debate, provoked much public concern. Possibly the dream had simply detected this and objectified the media images into a reality.

Also fascinating was the complex dream offered about the death of a celebrated British actor. I will not be more specific because, as I write, that actor is still alive. However, the death painted for him was said to be from an accident whilst filming and involved water. There was also to be a dispute over the will. This was registered on 2 September. A few days later, but not through the project, a second dream came to my attention from a man who clearly saw a different TV actor, Roy Kinnear, die in an accident whilst filming. She was so upset that she wrote to Kinnear, a letter he may never have received as, unknown to the dreamer, myself, or a third party this story was immediately relayed to, the actor was abroad filming a 'Three Musketeers' movie. On 20 September, during shooting, after crossing a river on a horse, Kinnear collapsed and fell from the beast dead. The same day as this story broke the news also carried reports about a dispute over the publication of the will of a third British comedy actor, Kenneth Williams.

Of the same order, but showing the distortion factor, was a very interesting dream registered on 30 August. This gave great details and a sketch of the dream scene where a 'jeep' or 'bus' had skidded into a lake or sea beside mountains and then sunk. Six 'survivors' were seen wrapped in blankets escaping the wreckage. In fact, 24 hours later an aircraft crashed into the watery approach to Hong Kong Airport, which in many ways fits the dream scenario. Remarkably, as TV images showed, bodies were wrapped in blankets. But there were six deaths and many survivors as opposed to six survivors and many deaths as this slightly off-key dream had contested.

Whether any of this material could have been used to predict any of these events is a highly debatable point. But in retrospect the percentage of interesting dreams was above what I suspect chance would predict.

Occasionally, these dreams of tomorrow can have strange consequences. I loved the one that Helen Thornton from Middlesex reported to me. She had a new boyfriend and dreamt that an unknown old man told her that he was two-timing with a woman called Cindy. Helen, for some reason, found this name silly and told her boyfriend this. Turning a little pale, he asked for a description of the Cindy in the dream. Then he sheepishly explained that he had been married to a woman called Cindy whom he had chosen not to mention before. As Helen phrased it: 'Our relationship ended.'

But it was rather more troublesome for Luton dreamer Chris Robinson, who was being studied by Dr Keith Hearne, a dream psychologist. He has had so many precognitions, notably of terrorist incidents, that he was once actually arrested by military police at an army camp after he told them in advance of a violent act that would befall them. Their only alternative to the idea that this man had really foreseen these details in a dream, which is what Chris Robinson argues, was that he must have had something to do with the terrorist operation in the first place! Of course, he quickly convinced them that this idea was nonsense and he joined the long list of psychics whose dreams have been so good they have brought suspicion upon themselves. In two other cases that I know about psychics were interviewed by police as possible murder suspects because their dreams revealed information that was never public knowledge!

I think the evidence that our dreams can see into the future is overwhelming. However, not everyone agrees. On 19 October 1983 I took part in a Granada TV programme discussing the theme. It was built around the ideas of Paul Barnett and Dr Peter Evans, two researchers who felt that precognition was pretty unlikely. The argument that billions of dreams being dreamt each night must by sheer chance produce some seemingly precognitive hits was the one they seemed to favour.

However, as fate was to decree that view was to be tested.

A member of the audience announced to the assembled gathering in the green room after the show was over that the night before he had had a dream in which a member of the cast of Granada TV's popular series 'Coronation Street' had died suddenly. This went down with the production team like the proverbial lead balloon, but must have registered in the minds of the sceptics amongst us.

Hours later one of the cast members (who played the character Bert Tildsley) did indeed die, suddenly and unexpectedly young from a heart attack. Millions of fans were devastated. I reminded Paul Barnett about this when we next met and he smiled: 'Well I never actually claimed that precognition does not exist!'

9

Predicting to Order

'I AM CONFIDENT there is no possibility of error when the prophecy is so unanimous ... Never again will England be involved in war.' So said Maurice Barbanell, editor of the prestigious journal of psychic and spiritualist matters, *Psychic News*. He was speaking on 22 July 1939.

He was by no means alone. As historian of psychic matters Kevin McClure has shown in his masterful research there were endless statements of peace all that summer long, culminating in the greatest folly of them all – the bald statement: 'As prophesied ... War does not come to your world.'

The very next day, Britain declared war on the Nazis and World War 2 began its six-year nightmare.

How can such a monumental event achieve such singularly dismal failures from the psychic community? Does it not alone demonstrate the invalidity of precognition?

Nor, of course, is it alone. In April 1992, still commenting on what the psychics say half a century later, *Psychic News* accurately reflected the cross-section of opinion from would-be predictors about the coming British election. Matching popular belief, these psychics were seemingly united in declaring a sweeping victory for the Labour party and the downfall of John Major's fledgling government. As could have been said – indeed as many like-minded mainstream media sources did say – one hardly needed to be in tune with the future to figure that one out. It was a foregone conclusion.

Only, it wasn't. The election confounded every political pundit, all the opinion poll surveys and most of the psychics. Not only did Labour fail, but John Major had a comfortable majority.

How can powers of prophecy be so inaccurate as to be worthless?

There is a difference. Probably a very important difference. The last chapter was a study of things that came 'in a flash', gut reactions, instincts, visions and dreams that struck from nowhere. What these many people describing their experiences all had in common was that they were reporting involuntary acts. None of them had sought the precognition. It just happened.

Whilst, of course, premonitions are not always right, probably because a lot of what we surmise to be precognition is really imagination, they are clearly more right, and right more often, than are the woeful attempts by 'do-it-to-order' prognosticators. That is almost certainly because glimpses into the future are occurring at the subconscious level, sneaking past the doorman in a variety of ways. They are just not amenable to being bullied into occurring as and when you need them, be it to see the winner of the 3.30 at Haydock Park or to reassure a frightened world that no war is imminent. If they were then there would be a lot of rich psychics out there!

There are a few famed psychics down through the ages whose work is legendary and who are widely believed to have 'seen the future' on a more regimented basis. Yet any examination of their success-to-failure ratio illustrates that we tend to view them with a blinkered perspective. We remember the occasional big hits and ignore the many times they were not quite so correct in what they told us.

Jeanne Dixon, for example, is an undoubtedly sincere and, at times, gifted psychic. Her prophecy of John Kennedy's rise to power and subsequent assassination, definitely documented in an American society magazine in 1956, years before these events occurred, makes that point. However, in being forced into the uncomfortable role of daily prophet she seems to have often issued statements that no stretching of the imagination could have labelled successful. For instance, she told us that the Soviets would be the first on the moon in 1965 (it was the Americans and four years later) and Richard Nixon would be a great president – which he may have become, but for Watergate!

Her comments were never difficult to find. During the 1960s and 1970s there were several books and biographies relating them. One summary gave a potted preview of her foretaste of the final quarter of the twentieth century. As this is drawing to its conclusion it is well worth reviewing her progress in that regard.

What was to follow? Well, the American two-party system (the Democrats and Republicans) would end by 1980. Despite a belated and soon aborted attempt by independent millionaire Ross Perot in

1992, it survived that election campaign still intact. Equally, without predicting the rise to power of Margaret Thatcher in 1979, we were advised that 'in the 1980s' there would be a female US president. Perhaps she meant Nancy Reagan?

The Vietnam war, almost over in 1972, was going to last for years and would draw the US into a terrible conflict with China in the 1980s. The USSR was after 'world domination' but a natural catastrophe would stop them in 1985. Many expected this to be a comet or asteroid crashing into the earth which Jeanne Dixon long claims to have seen and to know when it will wreak its immense destruction. Armageddon keeps getting pushed back – latest estimates being the mid-1990s – but Jeanne has promised to tell the world just beforehand – which seems rather odd. If she really believes this will occur and being the undeniably kind and Christian-spirited woman that I know her to be – is it not surely her duty to give the world the most information that she can so that we might prepare for this holocaust?

Of course, when the world is overwhelmed by this unstoppable natural disaster we can possibly all move to Atlantis, which according to another of the century's most revered predictors – Edgar Cayce (the sleeping prophet, as he was called) – should have already risen from the ocean bed in tandem with California, New York, Britain and Japan sinking beneath the waves.

However, and here is where the other problem with long-range prognostication comes into play, the effects of global warming, melting of the polar ice caps and the greenhouse effect, which nobody seems to have predicted in advance, will all very likely sink many coastal areas during the next 50–100 years. So do we say that Cayce was right after all – or, if we wait long enough, is something bound to happen that seems to match some prophecy that, logic really tells us, was actually designed to say something very different and which went quite unfulfilled?

A couple of interesting experiments have been tried to test the efficiency of long-range prophecy in this made-to-measure fashion.

American researcher Herbert Greenhouse gathered several psychics together in 1972 to get their collected prophecies for the next couple of decades. This provides us with a useful check on the capacity for prophecy made to order.

His 'Next thirty years: A sneak preview' (in *Premonitions*) is full of wonderful things – and with the proviso that a lot could happen in the last eight or nine of these thirty – it is not unreasonable to check on progress.

There was widespread accord that Japan would become the leading economic power in the world, with Germany dominating Europe. That was correct, but surely not very unexpected in 1972? Accurate again was the widely expressed view that Russia would become an ally of the USA – seemingly absurd at the time, but as events prove becoming more and more correct. There was little indication of the fall of Communism and the democratization of the east, probably the single most important historical event between 1972 and 1992. There was also widespread suggestions of a global war with China (with dates from 1979 to the mid-1990s offered). This looks equally absurd as I write.

Yet these were the only notable hits. Otherwise, the prophecies often look very silly today. According to these prophecies we should have cures for cancer, no world hunger, people living for a very long time and towns and cities under the ocean, a new form of theatre in which retired chat show host Johnny Carson will star as a commentator guiding people through a play, the discovery of a planet in space with trees, rivers and fruits on it. Not exactly spot on, or very likely to be so by 2002, I suspect.

The other experiment was in 1980 and had a more narrow focus. Seven noted American prophets of the mundane kind, i.e. they were less in need of having to preserve a reputation involuntarily thrust upon them, gave what they had dreamt or more often seen in visions for the 1980s and 1990s. This was part of a 'Book of Lists' survey of futurists, economists and writers who were suggesting how things would change in years to come. As such, one could actually compare the relative successes of the prophets who used only their knowledge and guessing abilities with the ones who professed some kind of added gift. I have to say that, whilst both were often wrong, the psychic prophets tended to be wrong more often than those who claimed to have no special powers.

We had, according to psychics, devastation in Chicago, or the Rocky Mountains, when a nuclear reactor set off a chain reaction between 1983 and 1985. Reprogramming DNA during the same period would end cancer. By 1984 offices will be working four ten-hour days a week to save energy. Three-dimensional TV will be in many homes by 1991 (or 1987 in another vision) and pollution will have been eliminated by the same time. The US currency was to be wiped out overnight in 1990, with ten old dollars becoming one new dollar. And (my favourite) in the 1980s people would start building moats round houses as a regular feature!

If you are so thrilled by the accuracy of these assorted predictions then watch out for treats still to come – notably the discovery of the secrets of Atlantis, apparently buried in a time capsule in an Egyptian pyramid, or the arrival of benevolent (and presumably stupid) aliens who consider us so advanced that they invite earth to join the galactic federation! I, for one, will not be holding my breath for these.

This is not to say there were no startling successes. No mundane prognosticator said how in the period 1982–3 Libya's Gaddafi would start to pose a real threat to the west, arming with Soviet weapons, or (a very near-miss about the wholly unforeseen Falklands War of spring 1982) that Argentina would dispute owner-ship of several small islands (sadly naming Chile, not Britain, as the ones they would dispute with). Or, indeed, that by 1984 Iran and Iraq would be at war with each other.

The above all came from a Texas woman called Bertie Catchings, which might make us want to take her predictions for other events more seriously than most. Unfortunately, the further ahead from 1980 that she went in her attempts to see tomorrow, the less precise became her achievements, although there were still some.

She saw cars being built with computer road maps inset into the dashboard and that by 1990 portable telephones would be carried in the pocket or handbag and be widespread – although not what a darned nuisance they would become to fellow travellers on the train or bus! However, I am rather more dubious of the (already two years late) underground railway tunnel between Chicago and Dallas that will carry 'bullet trains' at great speeds.

Mind you, the French have since 1980 built overground versions between Paris and Lyon which do much the same thing and by 1994 these are passing onward through a tunnel towards England (where they then slow to a crawl and must dribble into London).

Mrs Catching's extraterrestrial vision is more modest and feasi-ble too, being that in the mid-1990s a space mission will find min-ing tools left on the moon by visitors who came to our solar system millenia ago.

Moving ahead, by 2000 children will all wear electronic tags so they can never get lost, we will have an average life span of 150 thanks to medical progress and, rather less plausibly (indeed if this one hits it will prove precognition for all the world to see) that a sort of series of flexible bridges linking new towns or communities will be built across the Atlantic Ocean and this will allow people to walk from Europe to America!

Prophecy's greatest game, however (and indeed its industry), is the interpreting, and then reinterpreting, of the rhyming and jumbled verses of French mystic Michel de Nostradame. These were written over 400 years ago.

Nostradamus, as he is better known, was almost certainly penning his mystic visions about the century in which he lived, which is why he covered them in anagrams and symbolic references that made them almost impenetrable. He was clearly afraid of persecution from the religious and political figures he was discussing. If he was describing a war hundreds of years hence between two countries that did not even exist in the sixteenth century then why bother trying to be obscure? Who was going to understand – or care – if he told it like it was, or rather like it was going to be?

In any case, as we have seen, the strength and number of precognitions tails off markedly the further one gets in advance of the predicted event. Whilst rare images years ahead of time have been reliably reported, it is rather stretching credulity and flying in the face of all our knowledge about the way precognition works to believe that a detailed map of the future centuries hence could be sketched out by a French doctor.

The reason why the Nostradamus 'quatrains' or 'centuries' (as his sets of 100 four-line verses are variously called) continue to appeal to each new generation is their intellectual challenge. Aside from the Bible they form the only book that has never been out of print almost since the invention of printing.

They are, in effect, like a huge game show that anyone can take part in. Nobody has won, but the prize is a charted history of tomorrow. All you have to do is figure out how to decode these jumbled, apparently meaningless ramblings, written in obscure and archaic French and full of astrological allusions. Easier said than done – but great fun just to try.

By now some of his prophecies have been used half a dozen times over to 'fit' events in almost every one of the last few centuries. As soon as something happens that seems to be a better match, the previously accepted translation of what the prophet said is quietly replaced by a newer version – which awaits the next moment in history when it will be relegated to yet another 'past mistaken belief'.

That said, there are some intriguing phrases in the Nostradamus collection. Those wishing to study them will find Erica Cheetham's publications amongst the most learned and objective.

Typical of the prophecies are those that refer to Hister, such as Century 5 verse 29 which describes how 'The republic of Venice will be annoyed by Hister'. In Nostradamus's day Hister was an old name for part of the River Danube, so in all pre-twentieth century interpretations that was what it was assumed he referred to by this name. Then, along comes Adolf Hitler, and, as many of these verses seem to speak of a warlord, it is assumed that Nostradamus just misspelt Hitler slightly. No doubt in the year 2094, when the great worlord Bert Mister goes on the rampage, everyone will then point out that the H was really an M and everything will be squeezed into place again to fit a new set of events.

Come to think of it, when this does happen and ancient scribes dig up this book the prophecies of Jenny Randles will then be talked about because my – completely spurious – guess above just happened by chance to vaguely match something that sooner or later is bound to come close to being true.

On the other hand, what do we make of one of his last prophecies (if indeed the numbers mean much, which some doubt) – Century 10, verse 100 – in which, allowing for artistic licence and a few pinches of salt, he seems to say that 'England shall hold the great empire as all powerful for over 300 years – huge forces of power will travel by land and sea and the Portuguese will be a bit cheesed off'?

In the time of Nostradamus any expanding British empire was easy to foresee and as the Portuguese were, for a brief time, Britain's naval rivals, one can understand the quatrain, even if it means a lot less today. One can also juggle with figures and say that he was dead right, giving the length of the British empire mastered by sea power as 300 plus years (probably around 350 years most historians would suggest). So was this an amazing prediction or just an inference from logical deduction and a bit of luck?

It is not often that the prophecies are dated, but on the odd occasions that they are they create fewer grounds for reinterpretation and so are much more interesting. However, often the dates have to be worked out from the movements of the planets and can sometimes refer to periods of months or years or even several possible dates within each century. In only two or three verses are exact and specific dates given years ahead of their time. For obvious reasons these are by far the most important.

He does seem to have described a disaster in London in the 'year of six and twenties three' – which could mean 1666, or even 1966, of course, and if the latter and, assuming nothing like the great fire had happened on cue in 1666, could well have been reinterpreted to mean a Nostradamus prediction of the Aberfan disaster, on the premise (often applied to his words) that London did not mean London itself but the whole country it then ruled.

The real drama comes from applying these predictions to events yet to come. In keeping with just about every other long-range seer (not that this necessarily means much as we have seen) Nostradamus is widely tipped to have envisaged a global conflict at the very end of the present century (phrases like 'the great revolution of the numbers' being taken to indicate the imminent switch from the year 1999 to 2000).

Needless to say, knowing whether the verses you select to create your pastiche history of the future really do refer to this, not the past, or indeed just to pure imagination, is open to considerable doubt. But in one frightening quatrain (Century 10, verse 72) he is unusually exact that: 'In the year 1999 and seven months, a great king of terror will descend from the skies. This will revive the great king of the Mongols. Both before and afterwards Mars will reign merrily'.

It is not difficult to interpret this as a prediction of a terrible event in July 1999. But hang on a minute – should that be September 1999? It depends on whether you take the literal or astrological month – and indeed neither would apply, if you take into account that the calendar was changed after Nostradamus's time. Many days were missed out so as to put our calculated dates in line with what modern astronomers had decreed was the date according to the universe.

As for Mars reigning. Do we take this to be Mars, the god of war, meaning, as most have assumed, that a terrible war is already underway by 1999? Or, is it Mars in a more precise sense? If the Americans and the Russians launch a joint space mission to Mars and there is much speculation on what it will find during 1999 – by now a distinct possibility – do we change our minds about the meaning of the verse?

Yes, but what about reviving the mongols? Well, actually, Nostradamus wrote 'angolmois' – which is a word nobody understands, but some have suggested is a sort of anagram for Mongol rois – i.e. Mongol kings.

Of course, the thing descending from the sky is obvious enough, isn't it? A nuclear missile clearly. But what about Jeanne Dixon's comet coming to smite the earth. Say it lands in the northern hemisphere and devastates Europe and north America – possibly the ones to benefit most (i.e. be revived as new world leaders) could be the old mongol kings in the east. Indeed, in another, usually related quatrain (Century 2, verse 29), albeit linked for no really obvious reason with the 1990s, Nostradamus tells of how 'The oriental shall rise from his seat and cross Italy to see France . . . striking everyone with his rod' – so obviously, thanks to the comet, the eastern hordes will invade a demoralized Europe.

Or, as by 1993 the decoders of Nostradamus were already discovering, maybe these things are all actually previews of the suddenly arisen conflicts in a fractionalized eastern Europe, such as Yugoslavia. Talk is already rife of using the prophecies (including one about great heat cooking fish in the Aegean) as evidence of a forthcoming nuclear holocaust spreading from eastern Europe to Italy and France.

Or, to put it another way, pigs might fly (as indeed, oddly enough, Nostradamus also predicted when he wrote of 'The half man pig' and 'brute beasts' that 'speak' – which is widely regarded as a view of aerial combat and the masks that pilots wore during World War 2 that made them look a bit like pigs as they squeaked 'Red Leader – Bandits at one o'clock').

Indeed, just about the only thing we can really say that Nostradamus predicted through this very specific quatrain is that – sometime in later summer 1999 something pretty bad will happen. Not exactly the prophecy to end all prophecies that it looks to be at first glance.

I think the lesson of all this is that previewing the future is a very sporadic thing. It works best over the short term and in one-off dreams that tend to relate to things of personal relevance to oneself and to have strong emotional impact.

Occasionally, these dreams can span time to see more global events and, also occasionally, they jump tracks a lot more than a few hours or days. But these are exceptions. It seems very hard indeed to see the far future and, evidence suggests, when we do see it the outcome is less accurate than when it is connected with the near future.

As such, do-it-to-order prophets, are on to a loser from the start. This is not to say they can never have genuine insights. I suspect some of them do. But it does rather imply that we should be very careful how seriously we take the long term and more general advice that is the result of a production line of prophecies. It appears to be a production line fuelled by misperceptions.

10

Chain Reactions

WE DISCOVERED A PROBLEM when using precognition to try to foresee or to forewarn, even with modern computer technology. That problem was our difficulty in accessing sufficient data to make statistical correlations about trends or patterns within coming events.

To do that sort of thing we need more than one or two dreams that might describe a possibility. However, there could be a way to achieve that requirement thanks to something that I do not think anyone has written about before. I suspect this will prove a very significant new way of looking into the future.

I shall call this phenomenon the 'chain reaction' but do not profess that I discovered it. Indeed the idea came about in a way I will describe and, remarkable as it seems, I found that not only does it seem to work but it makes a fascinating kind of scientific sense that could have been anticipated from our pre-existent theories about time. The only real way to illustrate the concept is by way of an example. But first, we need to think about the whole idea of a coincidence.

Most of us assume from this word that we mean the chance coming together of two events which seem odd but are in no sense supernatural. You meet someone at a party and they tell you the date of their birthday. It is the same as your own. That is coincidence. You can work out the odds of it happening and they are not particularly high, because there are only 366 possibilities and there will probably be 20 or 30 of them represented at the party.

However, we also have an inner sense that there can be more to it than this. We all know the old saying that similar things, e.g. accidents, 'happen in threes'. This is a response to a subconscious recognition through the ages that the bunching of events is sometimes less random than it appears.

Earlier this century a bizarre relationship developed between a quantum physicist called Wolfgang Pauli, whose work on the nature of atomic reactions at the level where time and space behave mystically is still of great significance, and one of the best-known psychologists of modern times, Carl Jung. He had developed the concept of the collective unconscious – or that pool of information which at some root level we all share.

Physicists think this place is the location where quantum reality might be determined. Psychologists suspect it helps elucidate the patterns of behaviour which we all display. We, as students of time, should notice that it also seems to be that timeless, spaceless continuum that we found to be the source of time-travelling information. When data from here crosses the threshold past our doorman and reaches our individual subconscious mind a time anomaly seems to result – be it an experience of a past life, a telepathic rapport with a loved one or a precognitive dream about tomorrow.

We found that sometimes, when this information is not readily pushed into our actual conscious awareness, it can manipulate our behaviour on a subconscious level. As a result we have a gut feeling or intuition, may not board a train or plane, or, less directly, we call someone on the phone 'by impulse' at just the right moment when they need us, or step back from the window without knowing why before it is shattered into fragments by a rock.

These things are occurring on an individual level within all of us, possibly much of the time. Mostly we are not aware of them because if they motivate us to do something then we probably just assume that it came about by happenstance – as it probably does on some occasions.

But if, as physicists like Pauli argue, sub-atomic reactions are also to some degree conditioned by impulse choices and observation, shuffling the atomic pack to make certain combinations come into being, then we can extend the concept of the subconscious impulsive behaviour into the collective unconscious or, indeed, its control over sub-atomic reality itself.

As a result things may appear to occur within the world around us 'by chance' – i.e. they are just a product of the circumstantial atomic reactions and individual responses to them. But, assuming some kind of data transfer from the collective unconscious was actually involved, they could more correctly be a response to our collective detection of this unseen data. A bit like a gut feeling rippling not through one person's mind but the collective mind of humanity, or the quantum basis of sub-atomic reality.

Jung and Pauli, without really having the time-experience data to know what was happenlng, coined the term 'synchronicity' (i.e. two events coming together in time) for meaningful coincidence. This is an event that is more than just a statistical fluke, e.g. two people sharing the same birthday.

Here is a real case to help us see what is being proposed. A friend of mine was driving on a busy motorway when she suddenly ('for no reason', as she phrased it) swerved out from the lane that she was travelling in and entered the fast lane. She did not look. A speeding car could easily have been coming up behind her and the results would have been disastrous. There was no such car and she entered the lane safely. Then, perhaps ten seconds after her uncharacteristic motorway madness, the truck ahead blew a tyre and swerved to one side. Had she remained where she had been there is little doubt that she would have been hit. Because of her foolish act as a response to an unknown but subconscious impulse she saved herself from catastrophe.

From what we have discussed in this book so far you can probably see what must have happened here. Information about the coming accident rippled slightly ahead of itself through time. In other circumstances, if she had been asleep, it might have intruded into a dream and been visualized in some way either more or less like the real thing, to be remembered, or not remembered, as may be. Here that was never an option. The only way past the doorman was to tickle the subconscious into action – fast. So her behaviour was modified and she reacted 'on impulse' as a response to information that she was never consciously aware of having possessed ahead of time.

It is not difficult to move from this specific, individual case, to a broader view where the collective unconscious is the recipient of the same sort of advance information.

Let us say that it was the course that our society was taking that was 'risky', not just one woman's route driving on an unsuspectedly hazardous lane of motorway. Some danger lay ahead and the information about this could ripple through time. Whilst it may reach some individual minds to provoke some actual precognitions, such as dreams, it is very likely that it could also work at a collective level on this more subtle kind of behaviour modification approach.

James Lovelock conceived the Gaia hypothesis, arguing for the interconnectedness of living species and closed systems on earth, such as weather. The ideas of ecology and conservation draw considerably from this. But in July 1993 Lovelock offered a spectacular new idea as to why humans were so amazingly interested in dinosaurs. These beasts, which officially died out 65 million years ago – way before the first, even sub-human, mammalian creatures walked the earth have grabbed attention in an astonishing fashion. Steven Spielberg's movie *Jurassic Park*, built from Michael Crichton's novel about genetically engineering dinosaurs back to life, became the biggest box-office success of all time in summer 1993.

Some of this was hype. But people must have had a deep inner need to see dinosaurs survive on screen in order for it to work so well. Indeed interest in dinosaurs has been mounting for a long time. As such Lovelock ponders whether this is subconscious awareness by our species that we are approaching the end of our tenure on earth. If things progress as now the planet will soon go through an ecological crisis and humans will extinguish themselves. We have had nothing like the longevity of the dinosaurs, who ruled the earth for 100 million years and were far from the stupid, slow monsters most people think them to be. Yet they succumbed to some sort of huge natural disaster, widely thought to be akin to Jeanne Dixon's prophesied comet or asteroid strike,which changed the earth overnight.

If such a disaster were awaiting us just around the corner, then from what I propose above that event would without doubt ripple ahead through time far more than any other we have known. It would be bound to as it would affect far more people than even the greatest of mundane tragedies. As such we might expect to see it manifest in a variety of ways – e.g. a rise in prophecies of doom, greater interest in precognition, plagues of nightmares about climactic holocausts or fires falling from the sky. There may well also be more subtle effects – e.g. population movement trends, with people being subconsciously manipulated to move away from threatened areas to relatively safer high ground. These would be the signs to watch for that might signify that terrible things were coming swiftly upon us.

Lovelock, without being mystical or thinking along the above terms, says that our fascination with dinosaurs may come as a result of the detection below the threshold of our collective unconscious of that intangible and barely recognized sense that we are soon to share their catastrophic fate.

If this idea of subconscious detection at a cultural level is correct, we would expect to find several things occurring. People would be subconsciously motivated into acts that were odd and for which they had no explanation, indeed which might even look foolish.

But these would be in response to that hidden detection of information. On a general level whole changes in cultural trends may occur as reactions to such ripples through time.

Equally, and more contentiously, if actual events are the result of sub-atomic reactions (as most scientists agree) and if the outcome of these reactions are at least partly a consequence of humanity's collective consciousness (as many physicists are now being dragged towards accepting to explain their experiments into the sub-atomic structure of the universe) then there is an even more dramatic consequence. Any information detected subconsciously ahead of time at this collective level and which is powerful enough to create modified trends of behaviour, may also cause changes in the way reality itself will manifest. Since consciousness guides sub-atomic reactions towards specific outcomes, creating events and reality, then anything effecting changes at that collective level could just as readily adapt the outcome of sub-atomic reactions and, therefore, their appearance as specific events in the real world.

In other words, it is perfectly possible that real events could be manipulated into occurring by way of subtle, unseen but powerful forces within our collective unconscious. These would be further responses to information that has rippled ahead of itself through time.

What we would see would be a series of events that stressed a warning theme – just as changes in human behaviour may also be doing. Whilst we may recognize these behaviour changes as impulsive and not be too shocked by the thought of them being reactions to information received ahead of time, we would never anticipate a similar effect to be found within reality itself. But the above discussion strongly implies that we should expect it.

Most of the mysterious circles that have formed since the late 1970s in crop fields, particularly around revered sites such as Stonehenge in southern England, are without doubt the work of hoaxers and some are the product of rare natural energy forces. But they have still provoked an incredible social response without precedent. Millions have visited and marvelled. Psychics in particular have been attracted like bees to honey.

It is possible that the circles are an actual consequence of this detection ahead of time of the coming ecological disaster and a very visible means of expression of that warning trend. It would not really matter whether hoaxers were being manipulated into going out and spending summer after summer faking them, or weather patterns were being adapted so that they occurred more often and also in repetitive locations or whether they were unexplained rearrangements of the molecular structure in cells of the crops themselves. These are the three most popular rival theories, but if what I am saying is correct, then each would be anticipated as a consequence of the basic phenomenon that underlies the circles' appearance. In other words, everyone might have been right about circles after all – from the outright sceptics to the scientific experts and the 'crackpot' mystics. But what counts is not how something like this is happening – but why?

This would not be an isolated example. On a smaller scale, when dealing with dramatic but not world-shaking events like an ecological catastrophe, we should expect to see a series of happenings that look like coincidences (be they 'coming in threes' or whatever) and which are meaningful in their manifestation. These would be triggered into being by information rippling ahead of itself and shaping atomic reactions. The result would be actual events, like crop circles, not just dreams of them or modified behaviour patterns.

This offers a dramatic opportunity that we may have been ignoring. Warnings about the future may well be occurring on a worldwide scale in the most overt manner possible, where everyone can see them. But we are generally unaware of them because we are not looking for such trends or, when they do jump out at us from clusters of events, we dismiss them with the adage that they are just a coincidence.

Indeed they are a coincidence, but a meaningful one – synchronicity.

There are countless examples of synchronicity at work. I have a huge file, recorded from my own experience and from examples offered to me by others.

On a personal level one interesting, very simple case, came whilst I was sorting a huge pile of case files with witness names on them. As I was doing so far across the room my father had called home to my mother on the phone. I went through my pile transferring data and then got into a kind of mental block, holding one in my hands for ages, staring at it blankly, puzzling over the odd fact that it had the same witness name as the file I had just worked with. In truth, although I did not realize this, I was still holding the same file and behaving quite irrationally twiddling this in my hands for some minutes. There was no logical reason to spend so much time over such a minor matter. Then, suddenly, my mother called out to me – 'write down a name – your father wants me to remember it'. I had no need to write it down. The name was the one I held in my hand and had been juggling with for a pointless length of time.

I suspect what happened here was that my awareness ahead of

time of my father's need manipulated both the coincidence of holding the right file at the right time and the irrational subconscious behaviour in playing about with it for so long. Had my mother simply written the name down when it was said to her over the phone and not spoken her request to me, the likelihood is that I would have broken free of this spell and never realized that this moment of irrationality had any meaning behind it. As such, of course, it would have been quickly forgotten.

In another case, I was walking down a road in Seacombe on the Wirral where I then lived, desperately trying to bring to mind the quite unusual name of a previous doctor that I once had and now needed for some form. As I struggled I ambled across a road, without paying attention, and found myself going into the library. I had no need nor intention to do this, but had a very odd sensation that this impulse was important. So I let myself 'go with the flow'. I walked up to the counter and observed the queue. There was someone checking out a book. The author's name – plainly visible – was Pattinson. My doctor had, in fact, been Pattiniot. Needless to say the 'coincidence' immediately recalled that for me and resolved my problem.

The only thing wrong with such a description of that story is that I doubt very much this was an ordinary coincidence. I cannot convey the eerie way it felt as it happened and I suspect it was more a synchronicity.

A man from Liverpool told me of the extraordinary day he had spent fruitless hours searching the city centre stores for a small hammer that he needed for a certain type of household job. In sheer frustration he was wandering about aimlessly when he found himself in an alleyway behind two tall blocks of buildings. Suddenly there was a crash and he turned around to find a small toffee hammer, prone on the ground. He looked up but could see no sign of where it had fallen from, although there were open windows and the chances were that it had been accidentally dropped. The guilty party, possibly mindful of their dangerous act, had probably made a swift exit. In any case the man picked up the hammer and set off home – it was exactly what he was looking for.

Here, not only had his behaviour perhaps been motivated so that he was in the right out-of-the-way place to receive this windfall, but the reactions of the person holding the hammer may have been subconsciously affected so that he or she dropped it at just the right moment.

Oscar-winning actor Anthony Hopkins offers another example. He agreed to play a role in the movie *The Girl from Petrovka* but searched in vain through London bookstores for a copy of the book. Then, as he wandered into Leicester Square tube station, there was a copy of the book discarded on a bench. He picked it up, not even minding the writing in the margins.

Some weeks later when Hopkins flew off to make the movie he met the novel's author, who bemoaned the fact that he had lost his own copy after lending it to a friend in London. They had been unable to find it, but it had very important personal notes in the margin that were invaluable to him.

The book was, of course, the precise copy that the man chosen to star in the author's story had found 'by coincidence' on a tube station bench.

Clearly, here, the statistics are fantastic, given the number of copies of the book, the number of people who use the London underground each day and the odds that any one person would find any one book, let alone one they were desperate to find and, indeed, the only one that they would be in a near unique position to return to its rightful owner. These must be near enough to impossible and illustrate again that this has to be a motivated synchronicity.

There are hundreds of cases like this on file.

Now we are equipped to tackle the even more fantastic story I am about to relate. It crosses the boggle threshold by quite some margin. but I can back up much of it with hard evidence. I think it shows well enough that we should be on the lookout for chain reactions within the world at large.

We start with possibly the most disturbing of my own precognitive dreams.

In November 1983 I visited the USA and, thanks to my astrophysicist friend, the late Dr J. Allen Hynek, had dinner one day with a NASA astronaut. He was then planning the first mission, some two years hence, that would carry a civilian aboard the shuttle into space. We discussed this in some detail, such as who would be selected. As events were to transpire it was to be a school teacher, Christie McAuliffe.

A few days later I visited Cape Canaveral and was thrilled to see the launch of a shuttle mission. Both of these experiences were etched firmly on my mind as being memorable and exciting.

One of the main reasons I had been in the USA was to compile material for a book I was then writing on behalf of, and with, two colleagues, Brenda Butler and Dot Street (Dot travelled with me in the USA). This was about an incident that had occurred near a US Air Force base in East Anglia in 1980 for which no adequate explanation had (or indeed still has) been found. When the book was completed over the next three months I called it *Sky Crash*, a slightly odd title in retrospect, as it was about an unknown object that fell from the air and landed in the woods beside the base where senior airmen, reputedly, had a close encounter with it.

Much of the data gathered from the USA concerned a similar case in Texas that had occurred at around the same time and had some links, as did the British USAF base, with NASA. I titled the chapter, which was about an incident on a roadway involving three witnesses, the somewhat quirky 'Houston – We have a problem', mirroring a message from a space mission.

Neither of these titling decisions seemed odd at the time. They just felt appropriate. Only afterwards did I have cause to wonder.

Sky Crash was finally published in October 1984 and I spent a week visiting radio stations giving interviews. On the night before I went to the West Midlands I had a vivid dream. It was strong enough for me to record it briefly, not having time for a full account given my early start that day.

In my dream, I had seen the shuttle take-off and explode. I saw one of the chemical fuel tanks rupture and believed this was the cause. The effect was powerful, but I knew of no shuttle mission immediately imminent and, in any case, had no real opportunity to ponder whether I had visualized a coming NASA disaster, because within five or six hours of my dream it seemed to resolve itself.

That morning I set off from Cheshire for my first interview of the day in Wolverhampton but the train suddenly crawled to a standstill just outside the city. Some minutes later we moved on past a scene of devastation by the trackside, with fire engines and ambulances in attendance. When I reached my destination (Beacon Radio) I learnt from the lunch-time news that a huge chemical tank at a plant immediately adjacent to the railway line had ruptured and exploded in flames, killing one worker. This had happened minutes before my train was due to pass. I do not know if I was ever in any physical danger (although I was seated by the window on the side where the explosion did occur, perhaps 30 metres (100 feet) distant). In any case it was certainly the closest that I had ever come to such a disaster both in space and time and it seemed to make immediate sense of my dream.

I simply concluded that the impression of a chemical tank rupturing and exploding into a mass of flames was best dramatized in my mind by using the still vivid memories of the shuttle taking off – this being carried skyward on huge tanks of chemicals which are released through a controlled explosion. It was a typical, if harrowing, example of how precognition works inside your dreams, even if it had things wrong about it. For instance, the shuttle take-off was in the dark, but the mission I had watched (and this chemical tank explosion in Wolverhampton) had both been daytime episodes.

Just over a year later, on 28 January 1986, I still recall the moment when I was hurried away from my work to see the terrible scenes from Cape Canaveral where the shuttle had just exploded on take-off. It was the very mission that I had discussed two years before over dinner with that NASA astronaut – the first designed to take a civilian into orbit. As events were to prove it was also discovered that the catastrophe was caused by a fault in a ring linking the chemical fuel tanks to the shuttle. A rupture had created an explosion in the form of a huge fireball, just as in my dream.

Suddenly I had twin images embedded in my mind. The real scenes of horror screened again and again on TV as the shuttle relived its last few seconds. Plus my dream memory from 15 months before, which was very similar in many respects, yet different. The main difference – that again the fateful launch of the shuttle *Challenger* was in Florida daylight – seemed odd. But as I watched the TV images flickering before my eyes they were surrounded by a large bay window in my bungalow. That was a picture of blackness, because it was early on a winter's evening in England and so pitch dark.

Being personally wrapped up in this terrible event I probably saturated myself in the TV coverage more than I would have done otherwise. That, paradoxically, may have worked backwards through time increasing my chances of having foreseen this event.

Eventually I satisfied myself that the astronaut I had met in the USA was not aboard *Challenger*. In fact he changed his mission specification after we met and flew on a much later journey. Nevertheless, I was interested enough to look out for other people who might have had premonitions of the disaster. I made no special effort to trace them, such as placing advertisements in magazines or papers. But some stories gravitated naturally toward me as I expected them to do.

A typical example was from a Manchester dentist who had had a series of dreams in the weeks leading up to the explosion. He saw a strange aircraft fall to earth. Then he realized it was not taking off like a normal aircraft from a runway, but was rising from a near vertical position, but not exactly vertical. He was still wrestling with the confusing images when he saw the shuttle disaster and realized what had been in his dreams. The near, but not quite exactly, vertical angle of take-off is a combination of an effect you notice on TV images as the ground-based cameras point skyward to follow the rapidly rising rocket, plus the slight angle at which the shuttle does indeed take off.

However, the biggest shock came from a man called John Webb, who is a computer programmer from Warwickshire. Indeed, we began to correspond a month before the shuttle tragedy and were discussing our mutual interest in a story written by Nigel Kneale, entitled *Quatermass and the Pit*. This seems to have no possible relevance to the shuttle affair – but it does.

I recall seeing the BBC TV black and white serialization of this story as a child in about 1958. It scared the wits out of me. The story was remade into a Hammer horror film in 1968, which was virtually an exact duplicate but did not have quite the impact of the six-part BBC serial. The story is complex, clever and, whilst told in a delightfully scary science-fiction style, probably full of genuine insights into the paranormal, which is one reason why I had always been impressed by it.

In essence it tells of how a spaceship is excavated from a London underground station during digging work there. This is millions of years old and contains dead, insect-like creatures. Throughout history – from prehistoric to recent times – psychic events, such as visions of horned demons, have been recorded in this same location and Professor Quatermass figures out that the insects are our far distant ancestors who effectively engineered our society when we were sub-human Neanderthals. Deep inside we still contain race memories of these aliens buried in our subconscious and a combination of the crystalline metal of the spacecraft and outside forces, such as digging work, from time to time unleash these images from suitably receptive people. The story climaxes with a graphic image of the horned demon floating in a huge white cloud above the tube station.

Aside from this tale, John Webb had actually approached me to discuss a recently published book of my own in which I had written about an interesting synchronicity.

In his novels, *From the Earth to the Moon* and *Round the Moon*, Jules Verne had given a remarkable preview of the real disaster that struck the fated NASA mission, Apollo 13 some 100 years later. Verne, for instance, called his capsule the *Columbiad* and launched it from the coast of Florida, which was exactly what happened to the real Apollo 13 capsule that NASA had named *Columbia*.

Verne further described how an oxygen explosion prevented the crew landing on the moon and in a clever foretaste of real science returned to earth with a dangerous 'slingshot' tactic, the moon's gravity whipping the little capsule round its back and firing it on target towards the earth. It had to be an exact calculation otherwise the capsule would burn on re-entry through the intense friction of the earth's atmosphere. Luckily, Verne's crew make it home, splashing into the Pacific and being rescued by a ship.

Exactly this occurred to the real *Columbia* in April 1970. Effectively they relived the grand old writer's century-old fiction.

Possibly NASA should have suspected that someone was trying to tell them something, given their choice of name for the capsule plus the fact that astronomical calculations had decreed in advance that the launch of mission Apollo 13 would have to be at 13.13 on the 13th April!

Whilst Webb and I were discussing all of this in our letters Houston control at NASA were having one or two problems of their own. They had decided to name one of their three shuttles *Columbia*, which may not have been an altogether wise idea given the fate of Apollo 13. But during December 1985 and January 1986, the very weeks that John Webb and I were innocently discussing *Quatermass and the Pit* and the *Columbiad/Columbia* coincidences, a now legendary series of foul-ups hit the attempt to launch the latest *Columbia*. It was delayed and postponed more times than any other mission and the affair was much debated by the media. Several launches were aborted very close to take-off due to minor technical glitches that just kept cropping up. Eventually, after much effort, it was successfully launched.

Several of the aborted attempts were in adverse weather conditions akin to those that existed when *Challenger* followed a couple of weeks later. In retrospect NASA engineers discovered that the failure to the 'O' rings that bonded the chemical fuel tanks to the shuttle were compounded by the icy conditions. Had the minor glitches not prevented *Columbia* launching in such similar conditions it is perfectly possible that it would have suffered the terrible fate that soon befell *Challenger*. Today, the whole system is redesigned and the problem cannot recur, but *Columbia* had no such protection when chance intervened several times.

Oblivious to this chronicling of the traumas of the *Columbia* launch, John Webb and I began to talk about the movie *Superman III*. John introduced this because he told me that when it was first screened on British TV (5 September 1985) he had a 'bad feeling' about it, but did not know why. He was so bothered that he hung on to the relevant edition of *TV Times* magazine – the one that ran the premonitions bureau – see page 106! This displayed details of the cast list of the movie, etc.

When this news reached me a month before *Challenger's* launch date neither of us noticed the synchronicity – now so glaring within the pages of the *TV Times*. Adjacent to the screening – indeed going out at the same time on the same day – was a Channel 4 programme called 'Spaceflight' – a series of four films on that topic. This one was entitled 'The territory ahead' and discussed the future development of the shuttle and how, although we had become so accustomed to space travel that it was now considered as safe as taking a plane, it was virtually certain that a major disaster, probably to the shuttle, would occur and that this would be bound to affect the perception of society about this mode of transport.

Indeed this suggestion was to prove dramatically correct.

However, still neither John Webb nor I saw anything significant in this until after 28 January 1986. Why should we have done? John simply wanted to tell me about the feeling he had had about the movie and what (in isolation) seems a ludicrously silly importance placed upon this.

In *Superman III* the villain, Ross Webster, wants to control the world's economy by wrecking the coffee trade. He wishes to override a weather control satellite, Vulcan, named after the god of fire and from which the word volcano derives, of course, which will thus cause terrible floods in Colombia, South America, a leading coffee exporter. Superman must stop him.

In November 1985, two months after the movie was screened, a volcano in Colombia had really exploded, melting snow and, in a freak disaster, creating massive floods and heavy loss of life, plus destruction of crops. John Webb had wanted to point out that his bad feeling when the movie was aired, the links with volcanoes and Colombia in the story-line, and the real disaster that followed, may be somehow connected. He even added another seemingly feeble coincidence about the similarity of the villain's name in the movie (Ross Webster) to his own (Webb).

This does all seem petty in isolation. But what John Webb cannot possibly have known are the extraordinary links that also roped me into all of this. He had written to me simply because I wrote books, had expressed interest in synchronicities and written about the Jules Verne *Columbiad/Columbia* coincidences.

When I got his letter I had only just returned from Lanzarote in the Canary Islands. I visited here between the volcanic disaster in Colombia and John Webb first writing to me. I had been on a dormant volcano taking part in an experiment for a magazine (that had yet to be announced or published) in which I attempted to see whether both space and time could be bypassed through mental communication and readers could 'tune in' to where I was and what I was doing.

Equally, it is virtually certain that he could not have known that the movie *Superman III* had special significance for me because the villainess sidekick of Ross Webster was played by probably the only relative of mine ever to appear in a Hollywood saga. She is an admittedly somewhat distant one, as a cousin of my grandmother – but actress Annie Ross was something our family had often told me about. Even this seemed to bond tighter the feeble name coincidence John had mentioned – John Webb – Annie Ross – Ross Webster.

As January 1986 progressed John Webb sent me pages that he had secured from Nigel Kneale's *Quatermass* script and explained how the destruction of the Colombian town by the volcano had caused him to try to warn the Bishop of Coventry against a coal field that was due to be excavated. It was on a site that John was linking with the one in Kneale's story – called Hob Lane – indeed the same name as the tube station in the *Quatermass* story. The fact that 'hob' is an old name for the devil (as in 'hob goblin') was critical to the Nigel Kneale story-line.

Still we skirted round what we now see to be the real issues, as strange things went on at NASA and the endless delays holding up the launch of the shuttle *Columbia* continued.

In fact, NASA workers left in unprecedented numbers, some complaining about safety on the project, others offering no real reasons at all just leaving out of the blue. The media later remarked upon how this sudden exodus symbolized the end of the dream of space as a glorious thing to be involved with. Of course, in retrospect it is hard not to see this trend as rather like rats deserting a sinking ship, unaware why they were doing so but responding to inner urges that warned of the two-year cancellation of flights and heavy lay-offs that were only weeks away in the wake of the rapidly approaching *Challenger* catastrophe.

However, even as John Webb and I avoided seeing what is now more obvious, we kept digging up further clues. John pointed out to me that *Superman III* was made by Columbia Pictures. I could tell him in return that seven years before I had worked with that same company to help promote the Spielberg epic, *Close Encounters of the Third Kind*, and had even attended the première of that film. Its plot concerns a group of people motivated by subconscious urges that they fail to recognize to travel to the Devil's Tower in Wyoming where an alien craft is, unknown to them at any conscious level, due to land. Working with a movie company had been a unique event for me.

John bamboozled me further still by pointing out that *Columbia* derived from the Latin word for dove, that Dovemead was the name of the production company on *Superman III*, that doves were linked with the great flood in the Bible, which was even mentioned in the *Superman III* screen play, that Mary Magdalene in the Bible was 'the keeper of the doves' and that one of the causes of the destruction to the town of Armero, which the Colombian volcano had just obliterated, was its proximity to the River Magdelana!

We tossed coincidences like this back and forth. Of course, the difference from normal personal coincidence is that these involved real world events and seemingly immutable circumstances, such as film company names, stories in TV listings journals and natural disasters. We could not be accused of making any of this up or talking ourselves into it, because what we were discussing were objective facts, not subjective experiences.

As we still played name games and NASA struggled with its shuttles, I got another letter. This was written on 12 January 1986 by a man from Sussex whose name once again seems to have been a significant part of the sequence of events – Max Dangerfield! As he wrote, quite unknown to any of us, we certainly were approaching the point of Max(imum) danger! He had not read my book about the Jules Verne story but had read one of my UFO books and wanted to explain that on 9 November, a few days before the volcanic eruption in Colombia, he had been to a UFO lecture in London – at the Hotel Columbia! On its own such a trivial thing would have been unworthy of mention. However, it was by now far from being on its own. It was just another part of an escalating pattern.

Max also mentioned a curious dream in which he saw an outline of America on a map and an object climbing into the sky from its coastline. He assumed this was a dream of a UFO, but I sent a copy on to John Webb. The shuttle *Challenger* was still 14 days from climbing skyward from the coast of America, but *Columbia* was now doing just that. We still thought these dreams and synchronicities were all about the volcanic eruption in Colombia eight weeks earlier. None of us made the connection with NASA shuttles and the other *Columbia* then having such a struggle.

Eventually a third interesting story came from Bob Aldridge in Staffordshire. He was really upset because of what had happened to him. On 8 November 1985 he had seen in a dream clear liquid suddenly turning black with 'people struggling for their lives in the muddy water'. He saw a red glow, which he knew was a volcano, and then sang the words 'Armero' and 'Bogota', to his acute embarrassment.

Afterwards he consulted a map and discovered that Bogota was in Colombia but could not find Armero (being such a small place it was in few atlases). He did consider calling the Colombian embassy but could not link muddy water with a volcano and decided against this. Days later he was well aware of the tragic link between water and the volcano when the disaster unfolded.

Eventually, on 28 January, when *Challenger* faced its final moments all of these seemingly ridiculous things slotted into place in such an intricate lattice that it is almost incredible to behold.

My link with the shuttle, dream about the rupturing chemical tank, all the other shuttle dreams – these pointed the way through the most traditional of channels.

The more subtle motivational and behavioural factors had worked underneath the surface. Was I manipulated into chosing titles like *Sky Crash* and 'Houston – We have a problem' for the relevant sections of the appropriate book? What made all those NASA staff quit in the weeks leading up to the disaster? Was something deliberately fouling up the works trying to prevent the shuttle *Columbia* from launching throughout January 1986?

Yet behind such undercurrents, reacting to the ripples from the coming event, there were even more signs – things which manifested as a series of seemingly pointless, but in retrospect all too obvious, coincidences.

As *Columbia* was facing all of its difficulties, clue after clue was turning up via obscure real-world events that highlighted this one word, linking it with the volcanic disaster, even connecting it directly with spaceflight, the shuttle, 'the territory ahead' and a possible disaster. Several of us were in a position to have decoded all of this, but we were not even aware that anything was happening. We were not looking for it.

In Colombia the bottled-up fire from the volcano burst free, melted ice and created a disaster. The shuttle *Colombia* was saved from similar catastrophe by chance decree that it should not launch time and again, but eventually *Challenger* did go up and the fire contained within its tanks was released because of the ice on the rings that bound it together. The same factors – fire and ice – in all of these incidents.

As if to be the final, mocking irony, newspapers the world over published photographs on 29 January 1986 of the moment that *Challenger* exploded. This frozen moment of time from watching cameras down on the ground showed the billowing clouds of the fearful detonation. There was an image of the two boosters spiralling away at different angles from the main mushrooming cloud. It was even likened by commentators to a 'monster in the sky'. This was a very similar image to the horned demon cloud that dominated the final scenes of Neale's *Quatermass and the Pit!*

It is difficult to convey the importance of this, because it all sounds so trivial and incredible. But I think there are far too many incidents here not to regard this as a significant chain reaction of events.

I typed these words, at 8.30 pm on Friday, 16 July 1993, then I clicked off the computer and set off down stairs from my office. I switched on the TV set ready to watch a soap opera and relax, my mind swimming with these things and saying, 'Readers are never going to believe any of this. It all sounds so silly. Can I really be serious that it means something vital?'

I knew deep down that it did. I was well aware that this chain reaction was not the only one. American researcher Larry Arnold had made a similar study of the events on 28 March 1979 when the Three Mile Island nuclear reactor had malfunctioned.

Arnold's study noticed that there was the usual wave of personal precognitions in Pennsylvania of people foreseeing the accident – which became America's closest run thing so far to a radioactive holocaust. There were also the more subtle 'gut feelings', e.g. a plant worker who evacuated his family days before the accident and a pregnant woman who awoke the night beforehand, wandering around her house, restless, unable to sleep and with an urge to escape which she resisted. Arnold found that the frequency of these 'gut reactions' increased as the event neared, but he found some beginning as much as four years in advance.

Arnold found other odd things too. The machinery at Three Mile Island kept suffering minor glitches in the days before the near meltdown. There were so many they became noticeable. Also two other similar, but not as catastrophic, accidents befell other US plants in the two weeks surrounding the incident – supporting the motto about events 'in threes'. It was as if these were failed warnings engineered by our old friend – coincidence.

More subtly, Arnold traced literary coincidences, where fictional accounts seem to have described the coming events. The most remarkable of these is the hit movie, *The China Syndrome*, in which a nuclear reactor narrowly averts a meltdown in a chillingly similar fashion to that at Three Mile Island. It opened in cinemas at the nearby big town of Harrisburg, 12 days before the real accident. In the script, a character warning of the potential repercussions of the movie's fictional plant going critical says that an area the size of Pennsylvania would be at threat from radiation! Less than two weeks later this looked a lot less like just a throw away line in a movie and a lot more like an eerie prophecy.

It does seem ridiculous of me to suggest that links between things like coffee, Colombia/Columbia, volcanoes, ice and water are meaningful. But we see how the shuttle chain reaction was not alone. Another had been found independently. There must be more of them, but nobody is looking because nobody has realized that 'coincidental' events may be part of a pattern.

As I flicked on the TV set that Friday, doubts still simmered. But I was silenced in the most appropriate manner. Two adverts appeared before the soap opera. A stream of water flooded down a mountainside and a voice proclaimed the virtues of the coffee industry in Colombia. Then came a promotion for a programme on dinosaurs. An erupting volcano was prominent as its backdrop!

11

To Build a Time Machine

QUITE THE MOST RIDICULOUS story that I have ever heard was in Phoenix, Arizona, one persistently hot day in September 1989. For there I met a man who told me that he had travelled through time. Nor was this by way of any mental trick. He said that he had been catapulted 40 years into the future thanks to a time machine that had been built by the US Navy!

No doubt your response to such a bold statement will be the same as my own, to smile, be polite but consider the whole thing absurd. This opinion was to grow into near certainty when that man, Al Bielek, went on to explain that he had been born as Edward Cameron and gained a PhD in Physics from Harvard University. But after time travelling forward from World War 2 into the year 1983 he had been 'transferred' into the body of Al Bielek and made to forget the outcome of the secret experiment that he had witnessed. Bielek added that he only remembered these amazing details when watching a science-fiction movie called *The Philadelphia Experiment* and, as a result, understanding that what it described was based upon the truth!

Al Bielek appeared quite sincere and the story that he describes is on record as inspiration for the movie *The Philadelphia Experiment*. Its first telling dates back to 1955. Most have assumed it was nothing more than a crude hoax, but it remains a fixture in the minds of some researchers and has undergone something of a renaissance thanks to Bielek's amazing claims.

Given the allegation that a real time machine has already been built, we should examine what this American says and what can be established as the truth from out of his statement. Some of it is verifiable. The rest is a matter of personal choice.

The story goes back to Morris Jessup, one of America's earliest UFO writers, who received a rambling letter from one of his readers called Carlos Allende. He made various statements on subjects like levitation in the strident style that all recipients of such letters know only too well.

Jessup sent a reply but let the matter drop. Then, in January 1956, Allende further expressed his views that research into Einstein's long sought (and 40 years later still long sought) 'unified field theory' to link quantum mechanics and other natural energy fields into one unit had proved successful. In a test carried out by the US Navy the resulting energy had made a US naval ship and its crew become invisible. Upon their return many of these human guinea pigs were driven mad by their terrifying ordeal.

More material followed from the man, now signing himself just Carl Allen. Jessup would probably have dismissed all of this from further thought had it not been for the startling request to visit the Office of Naval Research (ONR) in Washington the following summer. Here the chief of staff, Admiral Furth, showed him an annotated copy of Jessup's own book – *The Case for the UFO* – which had lots of scribbles in the margin in a style akin to that of the Allende/Allen letters. Evidently these ideas had intrigued the naval officers rather more than they had the ufologist!

The writings roamed around many subjects and purported to be a three-way debate by persons noted as A, B and Jemi. They had passed the book and their comments to each other and often dropped heavy hints like 'if only he knew', implying that they were aware of things others were not. Above all they discussed two types of entities – the LMs and SMs – thought to be 'large men' and 'small men'. The latter were not especially well disposed towards humanity. There was also more about the disappearing ship, which supposedly teleported hundreds of miles and then back again to Philadelphia at the same time as becoming invisible.

In 1956 this commentary on two types of entities was all a bit of a mystery. There were still no such things as UFO abductions, as are prevalent in today's literature and folklore, even sightings of strange beings associated with UFOs were of relatively little credence in the eyes of most people. The concept of two entity types, both basically humanoid, but with the large ones more friendly and the smaller ones much less so is now firmly rooted within UFO history. However, that was not the case when this meeting took place in Washington at the ONR offices that summer.

Although it has always been stressed that this was a 'private interest' expressed by these US Navy officials, the calling of Jessup to the capital is a critical reason why the story of the Philadelphia experiment has not long since faded into history as an outrageous fraud.

In fact, the ONR took copies of the Allende letters sent to the writer and printed a small run of the annotated book plus these letters as appendices. In a brief introduction the ONR's special projects officer, Commander George Hoover, explained this was then to be distributed to various places for discussion, because of 'the importance which we attach to the possibility of discovering clues to the nature of gravity'.

All of this does rather imply, however sceptical we should quite properly be of the alleged US navy time machine story, that people at the naval research office had at least some idea that what Allende was discussing was possible. Of course, that is not the same as admitting that it had really occurred. All those involved at the ONR were later at pains to emphasize that no government money went into their interest or research and this follow through was strictly a private venture. None the less, such a depth of interest in what seems like a wild tale is still very puzzling.

In 1958 Jessup gave his friend, biologist and writer Ivan Sanderson, the original annotated copy and asked him to keep it safe in case anything should happen to him. Six months later Jessup was found dead in his car, an apparent suicide victim. There are those who feel that his death was more than that, but not a shred of evidence has been produced in affirmation.

Ten years later a man claiming to be Allen turned up at UFO group, APRO (the Aerial Phenomena Research Organization), admitting he was always just pretending to be 'Allende', had made all the annotations himself and sent them to the ONR to stimulate their research. He had made up the whole story. But he then later claimed that his confession was the hoax!

UFO researcher Jerome Clark notes that Carl Allen's parents have themselves labelled him a drifter and 'master leg-puller' and the view of most ufologists (pre-Bielek, at least) was that his claim is just fiction. Yet it provoked real fascination amongst naval research officers. Why?

Despite this seemingly fatal negative evidence, ufologist William Moore and famed mystery writer Charles Berlitz wrote an account of the matter in a 1979 book on the story.

There has also been an excellent science-fiction novel (*Thin Air*) written around the idea and in 1984 a movie, not of this novel, but supposedly built from the Moore/Berlitz reality. However, the movie soon wanders from the facts into Hollywood speculation. It was this film, which Bielek claims he saw in 1988, that reputedly triggered memory of having been a part of the real experiment in the Philadelphia breakwaters.

Bielek tells us that Nicolai Tesla, the renowned maverick electronics genius, was employed to develop various projects in the late 1930s that might have had offensive weaponry purposes. He reputedly built a 'death ray' (i.e. killer energy beams that knocked out power and could burn people). Both the Americans and British considered this weapon too awful to use – which seems difficult to imagine, given that these same sources five or six years later chose not to regard the nuclear bomb in the same context!

Tesla was then employed, along with other researchers such as mathematician Dr Von Neumann and with theoretical input by Albert Einstein, to build a device that could make a ship or plane invisible using quantum energy. Radar invisibility was sought, but optical invisibility turned out to result from the fields that were created. The project, as secret as that to build the atom bomb, was code named Rainbow. By 1941 it succeeded in its task using an unmanned ship.

Bielek (or rather Cameron, the physicist, as he then was, of course) was one of several naval juniors employed to work on the team and selected as the trial 'crew' when they first attempted to make a manned ship (the USS *Eldridge*) become invisible. However, Tesla fought against this use of humans in an untested project. He was reminded in no uncertain terms that they could not use experimental protocol in wartime and was made to comply.

That new test, conducted in late 1942, had failed. Bielek thinks Tesla sabotaged it. A few months later Tesla died quite suddenly. Another series of experiments were ordered for July and August 1943 and it was these that were said to have succeeded in the most horrific manner possible.

Reputedly the ship was made invisible, but still left an impression in the water. On the second test it vanished without trace. The warping of space and time sent the ship elsewhere and back again, returning with hardly anybody intact. Some of the men 'caught fire' because their atomic structure was out of alignment. Others were more fortunate and were merely driven insane.

The real horror involved those partly embedded within the bulkhead or the floor or the ship – their atoms melded into metal when still alive. Again this was due to the lack of correct atomic alignment.

Bielek escaped, along with at least one other, by jumping over the side of the ship which was 'suspended in a huge cloud'. Instead of landing in the sea off Philadelphia they were now at a research base on Long Island, New York. It was 1983 and the intervening 40 years had passed in an instant. They had also sucked a UFO through the time warp with them. It had been hovering above the ship, as had several other strange lights in this period between late July and early August 1943. It seemed as if it was aware in advance of what the researchers were up to and was keeping a watch on the experiment.

This seemingly ludicrous addition to an already ridiculous story could be very important – if we assume that the UFO was not an extraterrestrial craft, as most might anticipate, but came from a rather more earthly source. A source that knew in detail about the experiment afoot and might have the capacity to travel there and watch it as it was all unfolding.

Think about who – or what – such a source might have been.

It goes without saying that this whole story takes more than a little bit of swallowing! But there are some curious points about it.

Firstly, 39 years before the experiment – almost to the day – there are seemingly well-documented contemporary reports of a ship having a very strange experience off the coast of Philadelphia, exactly where the later experiment was subsequently reputed to have occurred. This event was in late July 1904 when the *Mohican* encountered a peculiar grey cloud that clung to its hull and created fearsome electrical effects. Compasses spun wildly around. Sailors' beards stuck out as if attracted by a powerful electrostatic force. Objects on board the vessel were magnetized. At the same time on shore there was a series of violent electrical storms and strange lights were seen to be drifting about.

If the Philadelphia experiment story is true, there was huge electrical activity in the atmosphere associated with a very odd cloud generated by the test. It sent someone about 40 years forward in time. That identical effects were reported in this area some 39 years beforehand is intriguing.

We have already seen several well-attested cases of cars and people supposedly teleported through time and space in the presence of a glowing cloud and strong electrical forces. There are further cases involving ships.

It has – not unreasonably – been suggested that the vanishing ship experiment was an invention by Carl Allen in 1955 based on his local knowledge of those fairly obscure 1904 newspaper references to the bizarre adventures of the *Mohican*. Whether he would have connected that seaman's tale with the simultaneous fierce electrical effects occurring on shore and also well attested at the time is more doubtful, as no contemporary media source appears to have done so. We can only say that this is possible.

However, there must also be a possibility that space-time distortions from an experiment in 1943 were somehow projected approximately 40 years forwards *and a similar period backwards* in time at this location. Does this hint that other cases of mysterious clouds we have met from all over the world and which have suddenly seemingly 'kidnapped' cars from roadways may, in fact, be further side effects from other real experiments occurring (or still to occur) during the quest to perfect a time machine?

After all, if time travel ever becomes a reality at any point one would not expect it to happen without some failures along the way. If the Philadelphia experiment is real, then could it have just been abandoned after the 1943 catastrophe? Bielek says it most definitely was not and work to perfect it goes on. Possibly its dramatic repercussions are causing the localized warpings of space and time that people occasionally report.

But why would these clouds of energy home in on ships and cars? Possibly these are bodies of travelling metal in an otherwise empty space and they attract the electromagnetic forces just like a lightning conductor attracts electrical energy. Any time-travelling cloud would seem to 'attack' a vehicle as if by some purposeful design. But, in fact, this would only be an accident of the laws of physics – not an alien invasion.

Scientists without ties to the military seem more willing than ever these days to support the idea that time travel is possible. Dr David Deutsch, a physicist from Oxford University, said in July 1992 that as yet 'we may not be able to build a time machine but at least we know it is not impossible'. He is not alone in seeking to do just that.

This level of conviction comes from our ever-deeper probing of the nature of sub-atomic matter. This book is not intended to delve fully into this, partly because it is very complex, but also because other books already exist which do a good job of expressing the wonders of this scientific advancement. I refer you to my reference section (see page 194).

However, it is useful to make a few general observations to understand what is now being turned from theory into practicality.

Before about 1915 scientists thought that matter could be broken into small pieces, called atoms, and that these were like building blocks. You had grains of sand, these formed beaches or sandstones and they in turn led to mountains. There was a finite limit to the smallness of all such particles.

However, the discovery of fields of energy that move at the speed of light throughout the universe introduced real problems. These radiating fields – such as x-rays, electromagnetic energies, various types of nuclear emissions and, most evidently, light itself – all behave very oddly.

Experiments showed that part of the time these fields, such as light, act as if they were composed of trillions of tiny particles, like atoms (in light we call them photons). You could fire a beam of light through a pinpoint hole and it would be like very tiny bullets passing through and hitting a target behind. Here the photons would be absorbed to create a patch of light. Fire more photons and the intensity becomes greater (i.e. the light seems brighter). Fire less and the reverse is true.

There is no problem here. The difficulty comes with two little holes through which the photons are directed. If both holes are open something curious occurs, the photons pass through each and hit the screen at the back. But there are not, as you might expect, two bright patches, one behind either hole. There is a series of bands of light and shade in a long horizontal line on the target area.

Such a band is also found with other forms of radiating energy fields that you can measure like this. The resulting unique fingerprint left behind tells you things about the radiating source and is a real boon to astronomers. They can use the bandings to help determine the make-up of a distant star just by measuring patterns from this radiation output.

However, the trouble is that this effect makes sense only if light or radiation does not comprise little pellets being fired out. Yet the outcome when there is only one hole (plus other experiments) prove that it is! To create the 'interference pattern', as these banding effects are called, you need to perceive light in seemingly contradictory terms, as a waveform, like ripples moving up and down on the sea. Reconciling whether light is made of particles or waves become a great dilemma for physicists.

It was effectively concluded that the question – is light comprised of tiny lumps or undulating energy waves? – is spurious. In truth it manifests as either form according to circumstances relative to how we perceive it, and our real-world language simply fails at this level.

The word 'relative' became a fashion when Einstein showed how time and space were inter-linked as a continuous whole (a space-time unit). Our perception of this was seen to depend on our relative location and dynamics. They were not rigidly fixed throughout the universe as had always been assumed.

But the fairly glib expressions of the particle/wave duality question, essentially saying 'that's how it is', masked the real problem. If energy fields can manifest either way dependent upon circumstances, then how does the photon or wave of light know what it is supposed to do?

It was like a game of cat-and-mouse. If you sealed off one hole and 'tested' the light it always responded perfectly as if it were little particles. If you opened up the other hole, it always reverted to displaying the properties of waves.

Never before in science did a demonstrable outcome seem to depend upon the actions of the experimenter which were fundamentally altering the nature of the thing that was being measured, indeed even according to a conscious choice that the experimenter could make, e.g. by changing their mind at the last minute and opening up the second hole.

It was almost as if the light beam was aware and could detect what decision you had made to ensure that it became either photons or a wave. Of course, this concept was clearly false. Science fought hard against it. But it was merely the first in a long list of ever-increasing clues that once you go smaller in size than the atom – i.e. into the sub-atomic (or quantum) world – things do not behave in any way like we have always come to expect them to, nor indeed how they seem to behave on the grand scale of objects much bigger than (but, of course, always built out of) these same atoms.

It was as if we crossed a threshold into another dimension where all the ground rules changed.

Einstein received his Nobel prize for work defining the earliest ideas of quantum physics, not for the less well understood (or at first poorly accepted) ideas of relativity with which he is now associated.

In 1932 physicist Werner Heisenberg also won the prize for his definitive proof that, disturbing as it may be to accept such crazy views of reality, accept them we must.

In fact, Heisenberg's so named 'uncertainty principle' went much further. He showed that, at the quantum level, you could never measure or predict anything for sure, just do so within a range of possibilities. If you tried to measure, for example, the speed of a sub-atomic particle (i.e. one of the many different things being discovered, and still being discovered today, which are smaller than an atom and part of its make-up), then you had to illuminate it with at least one photon of light just to theoretically observe it. Yet this photon will interact with the particle and alter its speed. Thus the observer becomes an integral part of the measuring process.

The whole idea of a detached scientist in a laboratory watching what happens, measuring things and then forming conclusions from the results may work when big structures like tables and chairs are involved. Once you are in the quantum world it becomes impossible. To a very real extent, the experimenter is an active part of the experiment that is being carried out, changing its outcome. In fact, there seems to be no fundamentally observable reality here. We can perceive only a distortion of it, forged to some considerable extent by our own interaction with, and our personal interpretation of, that underlying universal truth.

As the present century progressed we learnt more and more about this very weird sub-atomic world and each new discovery was like finding an alien planet with bizarre and unexpected inhabitants. In essence we have learnt that matter is comprised of energy fields that only sometimes take on an appearance of solidity and that many concepts of causality, time and space break down. Packets of energy called 'quanta' (from which the term quantum mechanics derives) make so-called 'quantum leaps' from one energy state into another and are in a constant flux. They do so by testing all future possible states that might be suitable and inhabiting every one of them simultaneously in a virtual, timeless state, before making any actual quantum leap into the one found most suitable or comfortable.

We have built particle accelerators to let these things circle round and round, picking up velocity, so that sub-atomic reactions can be made to occur at speeds close to light. From experimental results we have slowly been able to verify almost every important concept of quantum physics, despite most being as ridiculous as the time-related subjects that we are discussing in this book.

Neither precognition nor the Philadelphia experiment is more absurd – nor in any way contradictory to – what goes on in the heart of all matter. We live in a universe that is filled with time-travelling ghosts.

There are particles that approach one another, collide like snooker balls, move apart and split into other particles, which then collide – or rather start to – but then pass through one another as if neither existed at all. This is rather like a phantom Roman soldier walking through a cellar wall in York just as if it were not there. Elsewhere we have found hints of particles where time slows down or speeds up for them and even the elusive, but evidentially supported, tachyons that do the astonishing trick of appearing to travel backwards in time. We can now do experiments with many of these things.

Scientists who research quantum physics are like children in the presence of God. They know what they are witnessing is impossible and upturns all of their earlier beliefs and yet it is clearly true because all of their experiments prove that to be so.

Modern scientists often make statements such as 'anyone who is not shocked by quantum physics has not understood it' (Neils Bohr) or calling things that happen in experiments 'spooky action' (Albert Einstein). At times they sound like ancient mystics in their pronouncements, rather than hard-headed physicists. New physics is, in effect, endorsing very old claims about the interconnectedness of all things and the idea that solid matter and physical reality are an illusion. They are constructed out of an empty void filled with radiating energy fields that occasionally deign to confuse us by acting out the role of structured matter.

The bridge between this microscopic world of sub-atomic physics and the macroscopic one of tables, chairs, books and people is a time-travel gate between two parallel worlds where everything is different on either side.

We do understand to some extent why this change occurs. But this understanding has really only made things worse because it all boils down to statistics and coincidence – something few people like to trust. Indeed, so upset by this recognition was Albert Einstein that he felt sure we had done something desperately wrong and complained that 'God does not play dice'. He spent many fruitless years trying to find where he had gone wrong. Half a century later, we have to say, that God does play dice – and worse still – the dice can become invisible, travel through time and are controlled by our own consciousness!

The easiest way to understand this is to imagine tossing a coin. Throw it once and it lands either heads or tails. Throw it a million times and sooner or later it will land standing on its edge. Chance says that it must. Once you have tossed a head the next throw might seem to you to be a bit more likely to prove to be a tail, but that is not so. You cannot predict whether it will be head or tail. The odds always remain one in two because you have one result from two possible outcomes.

However, if you were to throw the coin 1000 times we can say that about 500 will be heads and 500 tails. The result will not be exact, but it will not be far away. Toss the coin a million times and the split will be almost exactly 50/50. This is something you can predict with certainty, even if it is completely impossible to predict the outcome of any single coin toss.

The same effect applies to science and explains why we have

happily gone through thousands of years of human experience without suspecting that anything is wrong with our knowledge. On the scale of the world of tables and chairs the numbers are so huge that the statistics almost allow predictability. Yet they are formed from a quantum world where nothing is predictable at all! The waves and particles inside the atom are statistical in nature. We cannot say how any individual one will behave. But even the tiniest grain of sand comprises billions upon billions of these, so we can make a (still approximate but reasonable) prediction of how it will behave.

However, there can never be absolute certainty. A chair does not move sideways on its own because there are so many sub-atomic things inside responding to outside forces like gravity that the odds say this is what it will not do. But it is possible that it would. It is arguable that the chair could move sideways. It is exceptionally unlikely to occur by chance – but, given the very long history of the universe, if the chair were always there, at some point it very probably would.

But there is worse to follow. These sub-atomic chance decisions can have utterly perplexing consequences.

The physicist Erwin Schroedinger proposed a thought experiment using a cat. But a human being works just as easily and shows the dilemma more graphically.

What happens is that you seal the person in a room with no ability to move themselves. Overhead is a canister containing instantly deadly nerve gas. One whiff and our volunteer, let's call him Fred, succumbs. The canister will open if a speck of radioactive material placed beside it emits a beam or particle in one direction. If that beam or particle goes in the other direction it misses the canister release trigger and this stays shut.

Now this is just like tossing a coin. We cannot predict what the next energy release will do. The odds are one in two that it will strike the canister and that Fred will die or that it will not and Fred will survive. But the big question is what happens next?

It seems a blindingly obvious experiment. When you open the door and look at the outcome after the predetermined interval, the speck has made one emission and Fred is either alive or dead, which is completely the result of chance – or as Einstein had it, God playing dice.

Unfortunately, that is not true. All the wealth of knowledge accrued from quantum physics tells us differently. We certainly are not misled by this knowledge. The mathematics is complex and yet well understood and to some degree experimentally verified. But what it tells us seems insane.

When we go away and leave Fred in that room, what is called a probability wave forms in space-time. Fred is in a limbo state, being effectively both alive and dead until something 'collapses the wave function', turning the assorted probabilities (in this case just two) into one reality and a now failed (i.e. collapsed) none reality. He is alive and is not dead, or exactly vice-versa. But what was it like for Fred's consciousness during this period when he was both?

Aside from the ridiculous idea of somebody being alive or dead at the same time, when all logic tells us Fred just sits in that room awaiting the dreaded moment when his fate is determined, there is an even more dramatic question that arises. What actually causes the wave function to collapse in the first place? In other words, what is it that makes the choice between life and death? Is it god or mathematical statistics? It seems neither!

In one sense it is merely a statistical quirk, but quantum physics insists upon a determining factor. There are really only two very different interpretations that physicists have been able to apply to this experiment, which has been to some degree successfully tried out (obviously without Fred, or, indeed, without Schroedinger's cat)!

In one solution championed by physicists and called the 'many worlds hypothesis' both possible solutions occur and reality splits into two new universes, in one of which Fred dies and in the other of which he survives.

As a result of the universe splintering like this at every occasion that a probability wave forms (i.e. trillions of times every millisecond) there are to all intents and purposes an infinite number of worlds side by side in which almost everything can happen, and, indeed, does happen. Our consciousness cuts a path through one of them – like a ship negotiating a channel through an ice floe without awareness of the mass of ice that surrounds it on either side. However, the uncertainty principle implies that our mind would be smeared across or straddle several adjacent channels where different splinters occur and differing alternative realities are found. We might be able to slip sideways through time by shunting across these.

This fits the evidence well, but seems too incredible for many scientists to accept simply because of the unbelievably fantastic number of side-by-side universes that would have been created since the beginning of the cosmos. It boggles the mind to even think about such virtual infinity.

Unhappily, the other solution is not an improvement. It might even be worse! This requires a hidden, as yet undetected, factor that collapses the wave function. For many physicists they can leave it at that, although it is, in truth, little more than a cop out – basically saying that 'we don't know what this missing factor might be, but why worry about it until we do'. However, there are an increasing number of physicists who believe that the missing factor is being well established by a number of experiments such as the following.

If you cause an atomic particle to split in a certain way it creates two similar emissions that are to some degree mirror opposites. They are, in effect, the positive and negative which when brought together make the more stable, neutral whole from which they originated.

These two emissions can shoot off across the universe in opposite directions, unconnected in any physical way. However, if you trap particle A and do something to it that changes its form, as well as measuring its parameters, then something very weird occurs. Particle B, wherever it is (or indeed whenever it is, because it may be moving fast enough to have bypassed our concepts of time), will always be found to possess the correct opposite stance should you also trap and measure that.

In other words, change particle A and particle B changes too, in exactly the correct fashion as if by magic. It does so if it is so far away that there can be no conceivable physical link between the two, or any known way for information to pass between them. It even does so if cause and effect are destroyed as a result – i.e. if particle B is bypassing time to such an extent that, in effect, it must change its state in anticipation of a change you have not yet made to particle A but will introduce!

The experiment was postulated in 1960 but it was a while before we could build the sort of equipment needed to make such difficult measurements. It was then conducted to some degree in 1974, but its most important reproduction was by Alain Aspect and his team in Paris in 1982 and it has been refined several times since.

It works every time. It does not matter how far apart the two emissions are, change one and the other particle now has the correct parameters to fit. If particle B has gone as far as the sun, then the experiment says it must instantaneously adapt its state. But as the fastest conceivable communication should be at the speed of light, then how can this be? The transfer of information must by normal physical laws take time, but, if what we are seeing is correct, it is instead quite timeless.

There seem only two ways of interpreting this shock result. Either there is some kind of communication on a mind-to-mind basis, or else our choice and determination of the status of particle

A is communicated by us in the same way to particle B and effects a change in that.

Of course, most scientists find it hard to envisage a particle communicating with another. So, those who are accepting this incredible outcome have been forced to conclude that it must be human consciousness that manipulates sub-atomic processes in some way. Our minds collapse the wave function. We are the missing variable in the equation that makes the choices that forge reality. As physicist Eddington said, 'the stuff of the universe is mind stuff'.

And, of course, if this is true, we are doing this by way of a consciousness field that spans the universe and is timeless in nature.

This last idea is a concept that we will not find hard to support through the evidence that has emerged in this book. Everything that we have seen – from the experiences of dreams to the nature of time-slips – has implied that at the heart of these things lies a timeless, spaceless consciousness field which links the universe into a single entity and bypasses all our indoctrinated perceptions about time. Or rather which shows that time is an artificial concept that we have imposed upon the world to help make sense of our confusing experience. At the fundamental level, where reality is first determined, time and space are seen to be just illusion.

The frightening thought from this, which physicists have not been slow to adopt, is that if we return to the experiment with Fred, it is our consciousness, our actual decision to observe the outcome, that somehow brings about the result. Fred dies or survives because of what we think.

Equally, all of this suddenly makes sense of pages 127–37, where I raised the seemingly novel and ridiculous issue of chain reactions. You must have wondered what possible scientific justification there could be for a series of seemingly trivial coincidences in apparently unrelated events somehow forming a pattern that was expressing a message ahead of time or even creating reality as it went along.

Yet, as we now see, at the level of quantum reality – the heart of all events in the world – outcomes are determined by statistical quirks which our consciousness does somehow affect. As such, the idea that information detected from a future point can ripple backwards and create clusters of real-world events (e.g. seeming coincidences) and effect behavioural changes within our subconscious to make people do things (from not catching a train to getting out of bed and taking the baby to safety or simply quitting NASA *en masse*) no longer looks fanciful. Indeed, it is probably a predictable outcome of quantum physics theory stating that consciousness is the 'hidden variable' that controls the outcome of statistical re-actions.

These are all useful things to remember when we look at how scientists are attempting to perfect a time-travel device. There are, in fact, three possible ways to develop the concept, each of which has a different outcome.

One researcher, Dr Frank Tippler from Tulane University in the USA has utilized the fact that in the concept of space-time as one unit all times exist at once. 'In physics the dinosaurs exist even though there have been none living for millions of years,' he points out.

On this premise it is possible to build a machine which can reach another time, in either the future or the past. These are constantly there along with what we call the present.

A simple analogy is to imagine a river. If you are on a boat this flows with the direction that the current happens to follow. If that is taking you toward some rapids you can do little to prevent that. Indeed you probably will not even be able to see them coming. At the last minute the water will get turbulent as the falls near and you can try to steer the boat away, but it will be far from easy and possibly too late.

However, if you can attach a motor to the boat then it can travel against the current and in order to escape all it has to do is produce more energy than that which is dragging you forward. If time takes us in one direction, with the current of the river, we have a limited ability to stop this. However, if we can create a motor that generates enough energy to counterbalance the force dragging us forward we can, in effect, move backwards from any coming event.

A better analogy would be to envisage being whisked from your boat by a helicopter and dangling on a line above the water. From this new vantage point you can see a limited distance back the way you have drifted and ahead in the direction that you are travelling. One assumes you are only seeing possibilities, not certainties. But you are perceiving the future.

You may see the rapids and can say with close assurance that the boat will reach them unless something is done to change its course. But you cannot be absolute. You may return to the boat and stop it before it reaches the rapids. By a conscious effort of will, fighting against the current and because of knowledge gained about the future, you will have changed things around. We have seen from real cases – such as the San Francisco woman who avoided the car accident (page 100) – that this appears to be what can occur.

Finding a device that can theoretically create more energy than that involved in dragging us inexorably forward through time, rather depends upon time having a one-directional flow. Quantum physics implies that might not be the case. But if it can be done, it probably means that a time machine would have to create an enormous power field and contain this within a small area. The attempt to build such a device into the structure of a ship is not unlike what would be needed. It would have to be that large. Even the choice of a ship is sensible if you think about it. There is a lot of empty space out there. Not only can you move the vessel without too much chance of being observed, but when you get where you are going you will almost certainly still arrive in an empty seascape without fear of observers.

If we are to take seriously the idea that a space-time warping field was possible half a century ago, then the only reasons why time-travel devices are not accepted fact today are these: either this theory is wrong and such devices are impossible, or, when built, the side effects on the space-time fabric and any participants were unforeseen and still uncontrollable. Exactly as Allende's story in 1955 first said they were.

However, physical travel like this is going to be much harder to achieve than we can imagine. Indeed, Dr Frank Tippler suggests that no machine could ever take anyone further back than the moment when it was first created. This seems to be the basis for the TV series 'Quantum Leap' and its intriguing speculation that mental time travel is feasible only within the boundaries of the lifetime of the traveller.

Rather more of a possibility is the traversing of time through some instrumental means. That is to build a device that can record sounds and pictures from times other than the present.

Indeed there are persistent claims that this has already occurred.

Alan Davies from Swansea told me that he fell asleep in his armchair one night and awoke with the TV set hissing static. He assumed that he had dozed off and the channel had ceased transmission for the night.

Yet the screen was not blank. On it was a fuzzy image of an animal. It was a dog that he recognized immediately as his own, as if captured on some freak home video. The only problems were that he had no video recorder and his dog had been dead for several months. As he got up to touch the screen, the image disappeared.

He interpreted this curious experience as being a possible vision projected from an afterlife. Most dubious sceptics would propose instead that it was a vivid dream, possibly a state known to psychologists as a false awakening. Alan only thought that he had woken up and seen the TV set. Instead he was still asleep. The sleepwalking act of getting out of the chair had then woken him for real, to discover, of course, that the TV set was full of static but there was no dead dog on it, yet assuming there just had been. He had no way to apply a reality check and prove that there had not been a dog on the screen.

I considered the possiblility of a scrambled TV transmission after hours, possibly conveying information when no programming is being issued by the broadcaster. But I could find no record of this.

We might contemplate the possibility that Alan had somehow experienced a vision. If a ghost of his dead dog can be perceived via his consciousness, as I think we have established, it matters little whether we call this a hallucination. That term has no unfair disadvantage of seeming to be derogatory. But to a very real degree all of our perception is an illusion created by mental processes.

If consciousness is some form of timeless, spaceless field, as many cases suggest it is, then there may be certain circumstances where we could objectify any perceptions through some other kind of electromagnetic field. This prospect is suggested by the video recording type of ghost experience (see pages 50–60) where both sound and images appear to get trapped in some stasis field within a building. We speculated then about rock and brick crystal structures triggering this field into being and the mind somehow decoding and replaying these. It is but a short step to imagine such a field being objectified onto a TV screen.

Possibly an interference effect could occur whereby the peaks and troughs of energy in one field can be made to resonate with those in another. This is rather how a sound wave, from a shrill voice for example, can set up vibrations in a distant and unconnected object, e.g. a crystal glass, transferring sufficient energy into this to break it apart.

There are some recorded instances of what is termed 'thoughtography', where a person seems able to enter an altered state of consciousness and transfer a mental image onto the chemicals on a film inside a camera. When these are developed they produce a fuzzy image of the thought picture. The most famous work of this type was carried out by psychologist Jules Eisenbud on American psychic Ted Serios. This research is somewhat controversial and relatively few successful pictures were produced.

154

But there have been anomalous images spontaneously trans-ferred onto film – such as Carlisle fireman Jim Templeton's photo of his young daughter holding a bunch of flowers on grassland in the Solway Firth. The unseen figure of a human-like being in an astronaut or radiation suit appeared on this when it was processed (see page 160 for more details). There are, of course, many other pictures that purport to show ghosts, and whilst most are undoubt-edly tricks of the light or hoaxes there are a few that appear to be unexplained. In these cases it may be that an image from another time which was in the form of an electromagnetic signal has been accidentally recorded by the film in some way.

More interesting still is research detailed by Florida psychiatrist Dr Bethold Schwarz, who studied a woman called Stella Lansing in Michigan. She seemed able to transfer by thoughtography images from her mind directly onto both still and movie film. Several anomalous films do exist in which thought constructions appar-ently were recorded for posterity in this way.

There were also some very curious anomalies in this research where strange human-like entities were discovered on the film, as if walking around the area and yet completely unseen. It is a bit like imagining them to be figures trapped in another dimension slightly out of phase with our own and somehow the mind of Stella Lansing has 'resonated' and transferred their images onto another type of recording medium – the film.

There is also a series of contentious experiments into what is called the EVP (electronic voice phenomenon). These began with the work of a Swedish man, Friedrich Jurgenson. In 1959, whilst recording birdsongs on sensitive equipment, he picked up unexpected voices etched into the electromagnetic field of the tape. They appeared to be conscious entities discussing the work that he was doing. But nobody had been visibly present anywhere near his location.

Much more sophisticated equipment was eventually developed by others to overcome a reasonable sceptical argument, that these were stray radio transmissions being tuned into and recorded. The voices sometimes seemed aware of the experiments and made intelligent comments.

In 1977, George Meek, a rich American inventor, perfected this equipment to such an extent that two-way contact with the voices became possible. The so-called Spiricom device was said to be a link between this world and the afterlife. A physicist who died some years ago supposedly worked on the project once the link was set up – so Meek and his team report!

Some people find these claims too amazing to accept, although duplication experiments have occurred with a limited degree of success. The main problem in listening to the tapes of these conversations is getting used to the robotic feel that they have. This is supposedly a consequence of the way the system works and once you are familiar with that the conversations are easy to follow. Reputedly the equipment issues an electromagnetic signal and the 'unseen entity' from another time (or the timeless afterlife, if you prefer) has to modulate this with an electronic carrier beam. The resulting interference effect manifests as a simulated voice.

There is little doubt in my mind that there are only two answers to this work. It is either an outrageous hoax, which its main protagonists insist it is not, and it seems to lack financial motivation at least. Or it is what it appears to be – communication across time and space with what we would call 'ghosts'. There is little scope for any middle ground.

Further work in this area goes on and in mid-1993 a group in Germany announced that it had made the first real breakthrough in attempting to capture these images across time and space directly onto video. This would seem no more unlikely than audio recordings and, whilst the first results are poor and in need of much verification and duplication, the possibility that this kind of process could occur is apparently increasing.

Interestingly this work coincides with two reports made to me – one from south Lancashire and one from south Wales (oddly not far from the scene of the 'dead dog on the TV screen' case). In both these I was told that the person observed a strange set of images and/or static on TV and a voice announcing that it was a test transmission from the future.

My immediate thought, and indeed that of at least one of the witnesses, when these things occurred in late 1991 was to assume that this was a new advertising campaign to promote some technological item. In 1993 the Sega games company adopted advertising where they simulated a pirate TV station breaking into transmissions for the duration of the promotion to sell their electronic equipment and there have been other attempts on similar themes, for example, pretending that a news flash is being interrupted or even that aliens have taken over the TV station! So the idea of a campaign using what purports to be a TV signal beamed from the future telling people to buy some product looks quite within the realms of advertising gimmickry.

However, I could find no such advertising campaign or see one myself, when I watched for it. Two years later, the advertisements seem not to have been repeated and nobody else I have asked has seen them. From what I can gather the effect was just as if some signal had overriden the normal TV picture and very briefly, via some rather meaningless imagery that promoted nothing, stated that it was an experimental transmission across time using TV signals.

If anybody else has seen such a thing – or indeed can explain what this was more simply – do write and tell me. I suppose it is possible that an experimental technique from the future could create interference effects like this today. In that sense we would be in a position rather like those of the California scientists in physicist Gregory Benford's novel *Timescape* who receive the tachyon signals 30 years before they are transmitted on their way back through time and are, of course, mystified by them.

However, as we have found so far, the simplest way to travel through time appears to be through that timeless, spaceless consciousness field. So probably the easiest type of time machine to develop would be one that finds a way to make such mental time trips work to order, rather than infrequently and by accident. We probably require either a chemical or electrochemical device which can adapt our state of consciousness and can urge us to be propelled forward or backward in time.

In a sense, regression hypnosis and creative visualization, the sort of techniques used to guide people into 'past-life' experience, may well be doing this. They possibly instil changes in the level of consciousness that could be measurable. Hormone secretions may alter or there will be real electrical changes taking place in the activity within our brain.

Other than putting someone under a CAT scan device to measure brain activity whilst having any form of time-slip experience we will only have trial and error to test what kind of changes are most helpful. Unfortunately that sounds like an appeal to legitimize all kinds of mind-altering drugs and free experimentation with these. Some such work, under carefully controlled conditions in a laboratory, might be useful, but generally such a ploy could be dangerous as the effects would be unpredictable. Sadly, the use of a CAT scan is also unlikely because these are very expensive devices needed for urgent medical purposes and no hospital (quite rightly) is going to sacrifice that for the sake of weird time-travel experiments.

Nevertheless, at least one researcher claims to have already succeeded in creating a machine that will stimulate time travel. The man, Tony Bassett, says that he built it under a railway arch in Chalk Farm, north London, as a by-product of his research into the way healers can emit energy waves that are absorbed by the body and can improve health.

People who have tried his device, which is about the size of a shoe box and is placed right next to the head as the person lies down, say that it creates a tingling sensation and a feeling of dual consciousness. They are aware of being rooted in this world but find their mind drifting upwards and away like a tethered balloon. They can return to base at any time but explore wherever and whenever in the meantime.

In experiments so far, number plates on vehicles parked some distance away have been read and then confirmed to be accurate, suggesting, Bassett thinks, that the perceptions of travelling to other locations and times may also be correct and not just a fantasy. Most sessions have been to the past – the good percipients having a full sensory perception of 'inhabiting the body' of another person in that time, seeing what they see, feeling what they feel and so on. Some experiments into the future suggest this may also be possible and Bassett reports that when 'drifting' people seem instinctively to know how to reach future or past almost as if it is a spatial direction.

The device reputedly works by creating effects like those in a strong electrical storm, both ionizing and electrostatic radiation waves that are transferred directly into the cortex of the brain. This presumably stimulates electro-chemical changes in the brain which trigger an altered state of consciousness.

That is fascinating, simply because it is exactly the kind of energy that appears to be detected by witnesses to time anomalies – such as those floating clouds, and, indeed, in a possibly not unrelated observation, by witnesses to UFO close encounters. These people often talk of tingling sensations, hair standing on end, eyes watering and sensory disorientation, which suggests exposure to large levels of electrostatic and ionizing radiation fields.

Possibly all that Tony Bassett has done is make a machine that duplicates these naturally occurring energy fields that trigger time anomalies. It seems that in its present form Bassett's machine still needs a 'guide' to stop the person drifting completely away, probably into deep sleep, and to keep them fixated on a time-travel task. However, presumably this could eventually be automated.

Whether any, or all, of these methods of time travel, or any of these devices reputedly built to circumnavigate the dimensions of space and time, will ever become a commercial proposition remains to be seen. However, one basic fact is clear. Time travel is moving out of the realms of science fiction towards the day when it becomes science fact.

It is, as they say, just a matter of time.

12

Have Time Travellers Landed?

WHATEVER THE STATUS of the Philadelphia experiment there is a vital point that is gruesomely illustrated by it. All devices meant to carry a person through time, from those in H.G. Wells's 1895 story to Ian Watson's modern day *The Very Slow Time Machine,* are designed to stay in one location which is on the ground. This is usually within the scientist's laboratory. However, that is simply not possible. Indeed it would be downright suicidal.

We noticed from cases where ghostly images are seen as video records that really old phantoms, such as the Roman soldiers in York, seem to walk with their feet partly embedded beneath the ground. We argued that this was because such temporal records follow the floor level as it was 2000 years ago, when that signal was somehow 'etched' into the location as a form of electrical field. The floor level has raised over the centuries by several feet thanks to building work and geological factors. So in our time what for that era was ground level is now below our feet. The further through time you travel the worse this displacement becomes, but it is always a factor.

Suppose you travel from a point in your lab to last Wednesday night but forget that you then had a table on the exact spot where your time machine now sits. Perhaps you assembled the final parts of your device on it. When you materialize seven days ago you are at very real risk of doing so slap-bang in the midst of that physical object – just as unlucky sailors standing in the wrong positions on the USS *Eldridge* were said to have done after their trip through space-time in the Philadelphia experiment.

If you travel a hundred years forward the building that you now work in might by then have been demolished and a busy motorway might run through the position. Your time machine could then appear right in front of a gigantic truck, merrily rocketing along oblivious to your arrival, which flattens you.

Any journey backwards by thousands or millions of years could see the spatial location where you have begun your trip hugely altered, submerged under an ocean or raised up inside a volcano, for example. There is just no way that any machine developed for a physical move through time could do so whilst remaining on the ground.

Indeed, there is only one relatively safe place where such a time transfer could happen – that is well above the surface, or better still outside our atmosphere altogether. Even that would not be completely secure from meteors or a chance meeting with a comet during its orbit, although much could be avoided by calculating precisely 'when' you were going to materialize. In any case, the mathematical risk would be minimal and you could be reasonably assured of not materializing inside something solid. That would definitely not be the case if your time machine remained on the ground.

It is now worth returning to the Jim Templeton photograph mentioned in the last chapter. Recall that he took a picture of his daughter in May 1964 and on development in the background appeared a strange person dressed in a white suit. Nobody was visible to the photographer on this remote marsh land area. Having talked with Jim I am as sure as I can be there is no trickery involved. Kodak, who analysed the negative, ruled out double exposure (i.e. superimposing one picture onto another by accident). Indeed, the company set up a (still unclaimed) reward for anyone who had an answer to the picture.

This image has been variously ascribed to aliens otherwise invisible to the naked eye, a picture of a worker at a nearby nuclear plant imprinted on film in some thoughtographic way, or even a ghost. But there is one important fact that immediately stands out when you look at this photograph. The figure is standing in a peculiar way. Whilst we cannot see the bottom of his legs, we can tell that if we could do so then he would be several feet above ground level – literally floating above the surface.

Now, think about a time traveller returning into the past and finding him- or herself in a completely new environment. Just as the past ground level of, say, 500 or 1000 years would be below ours today and an image from that time projected forward into our present would thus appear to be partly below our feet (as indeed some real ghost cases are), the reverse would also apply. Go back to the day when that Roman troop left its imprint at York and we (by standing on today's higher ground) would float above *its* floor surface!

Therefore, any person whose image returns backwards from the future would appear within our present time at a level consistent with wherever the ground 'will' be at that future point. This may well be some feet above our own present level. Any figure from that future time briefly appearing in, let us say 1964, and captured on film, would seem to us to float a little up in the air – exactly as on the Jim Templeton photograph. Is the image on that picture the future equivalent of a ghost photograph – a phantom person from a few centuries hence, perhaps?

This possibility is increased by the fact that several witnesses who claim to see ghosts on a regular basis report such an anomaly. One teenage girl who had a track record of countless strange experiences, including seeing aliens and UFOs, having precognitions and one time-slip when she saw a rerun of an aircraft crashing in World War 2, told me how when she was four or five she used to see people walking about the streets in central Liverpool, some partly within the ground and some floating a few inches above the pavement. Other than this they looked completely normal and she 'grew out of this' strange perception only when it became clear that nobody else was seeing them. So gradually she stopped doing so herself and confined what she 'saw' to what was 'normal'. It was as if she was tuned into past and future perceptions at the same time as the present, but trained her doorman to block out these anomalies from her mind when she recognized their unacceptability to others. In effect she adopted our view of consensus reality.

Interestingly, many reports of what are purported to be alien beings – although often so evaluated for no better reason than that they are strange human-like figures of unfamiliar appearance – are commonly reported by witnesses to be hovering just above ground level. The literature of such cases is voluminous and in a survey that I carried out almost 20 per cent of 500 reports referred to this otherwise startling and somewhat baffling feature.

Why should aliens hover a foot or so in mid-air – or, worse still, on occasion, walk normally 'but as if their feet were suspended on a cushion of air' (to paraphrase independent witnesses in cases from Australia, Argentina and Sheffield in Yorkshire, all found at random)? Of course, if these are images coming from the future of our own world, not the arrival of visitors from distant stars, then this problem would be eliminated.

Because of all of this, a time-travelling machine would need to avoid shifting time zones in close proximity to the earth's surface.

It would need to be designed to climb skyward, perhaps into orbit, make the transfer there and then return back to the surface again something like a space shuttle. So, if we search historical records to find evidence of such a device having ever visited us in our past and coming from some point into our future when time exploration has become feasible we should not expect to discover an H.G. Wells-type machine trundling along the ground. In fact we are looking for something quite explicit: unidentified objects that appear to be advanced flying machines cruising through the air, appearing suddenly, coming to earth, staying a brief while, climbing upwards again and seeming to vanish quite suddenly back into nowhere. As you might have surmised from this we are looking for something that we just might choose to call – indeed *actually have* already called – UFOs!

This interpretation of UFO activity as time travellers is little discussed by its researchers. The mind set of ufologists that these reports stem from extraterrestrials visiting the earth from a distant star system is so engrained it pays little heed to the occasional articles on the theme that have been written, such as those by Dr Michael Swords, a professor of Natural Science at Western Michigan University. As Swords points out, our knowledge of space and time no longer preclude time travel. Indeed the discovery of astrophysical phenomena such as 'black holes' in deep space even seems to offer hard evidence that 'time gates' exist in nature. These holes are black because space-time is warped to such a huge extent that even light cannot travel fast enough to escape.

In addition, we might note, there seems to be an implicit vindication of that timeless nature of consciousness that we found earlier in this book. Einstein argued that nothing could travel at the speed of light because here the passage of time contracted to zero and any material object as a result became reduced in length to nothing. Obviously this is nonsense. Such an object would have to be everywhere and nowhere at once (or indeed everywhen at once). The equations designed by the physicist Hendrik Lorentz, prove this beyond any acceptable doubt and so the idea of any speed of light travel is effectively eliminated by such nonsensical repercussions.

Or are they nonsensical? We know that something *does* travel at light speed – light – indeed any type of electromagnetic radiation does just that. If consciousness is established as an energy field like this, then it too would, by necessity (and relatively speaking), possess both zero length and zero time – i.e. be timeless and spaceless. This is exactly what we have argued throughout this book so as to explain phenomena like precognition.

The way physics handles this dilemma is to point out that radiation fields are non-material. What Einstein and Lorentz really established was that no material object, like a metal spaceship, could travel at the speed of light, because in doing so it would perforce become immaterial – in effect a radiation energy wave pattern. It would vanish from our normal perception just as if the book you now read were suddenly transformed into nothing but an electromagnetic wave. We would be able to detect it, with the right instruments, but it would not be visible to our senses as a book. Nor would it be in any sense restricted by all those concepts of space and time. Of course, as we noticed about the nature of all quantum mechanical particles – the building blocks of the universe – they do seem to have this weird ability to be solid or wave form in nature according to circumstance.

As such, for a device to travel through time you would not only expect it to fly like a UFO but you would equally expect it to do amazing things that visiting spacecraft would have no need to do. If our rockets go to Mars and start exploration they will not instantly appear in the Martian atmosphere and vanish again on departure. But if an object is cheating time that is exactly what it would do – by phasing between a material object and a radiated energy – just as, indeed, do sub-atomic features like photons.

Consequently, the image of time ships rather than space ships flitting around our atmosphere fits the way in which UFOs behave remarkably well. For it is a consistent feature of these sightings that they appear suddenly and vanish just as rapidly. Indeed, astronomer, Dr J. Allen Hynek, who was a doyen of the field and whose work inspired the Spielberg epic *Close Encounters of the Third Kind*, concluded that the idea of space visitors in the usual sense did not work for just that reason, plus the problem of there being too many UFOs coming here. Any alien race sending them to us was going to absurd lengths to precipitate the thousands of seemingly bona fide observations.

Hynek further noted that, unlike a jumbo jet, which took off from an airport and landed at another one and also could be seen at many points along the way, UFOs circumvent space. They do. They also seem to bypass time.

In passing we might also note why, assuming Al Bielek is sincere, the idea of a UFO observing the Philadelphia experiment is hardly ridiculous. In fact, if UFOs are time ships from the future and that experiment was perhaps our first serious attempt to create a space-time warp, leading many years later perhaps to controllable time travel, then future voyagers would know that fact and be intensely

163

fascinated by it. What aviator would not want to witness mankind's first flight if that ever became possible? What time traveller would not want to be there when the first time machine took off?

Another critical problem with the UFO evidence is the remarkable durability of these sightings. This is partly explained, as any good book on the subject will tell you, by recognizing that 95 per cent of UFO sightings are, in fact, not UFOs at all. They are what we term IFOs – *identified* flying objects, being anything from aircraft to balloons or real peculiarities like owls that have eaten diseased fungi and are glowing in the dark!

Equally, even the unresolved sightings often describe phenomena that seem to be natural effects that science has yet to catch up with: ionized plasmas, unusual forms of ball lightning, geophysically created lights squeezed out by processes within the rocks and so forth. They are lumped together under the umbrella term – UFO – simply because we have a desire for order and pattern and a need for one simple answer to any mystery. But, there is clearly no unified theory of UFOs. They are a wide variety of explained and unexplained events coming together to produce a rag-bag assortment of sightings. That said, there is a category of UFO that still generates perhaps 2 per cent of the data (which in total represents perhaps 10,000 reports worldwide each year). These few cases are consistent and almost impossible to account for as misperceptions or natural phenomena. Two per cent of 10,000 reports still means several hundred 'visitations' a year by what appear to be stranger craft with magical properties that defy both space and time.

What is even more important is that these 'devices' are not confined to modern times. They were reported throughout history. You see hints of them, e.g. as 'heavenly machines' in the Bible, or they appear as 'flying shields' in Roman times and 'marvellous airships' in the last century. If UFOs were spacecraft we have to explain why they would appear to have traversed this world for tens of thousands of years. Why stay so long?

These concerns will all evaporate if we assume the visitors are coming here in time ships from some point in our future not in space ships from another galaxy. It stands to reason that time travellers would appear to us in the same basic form throughout history, being naturally mistaken by the culture of the era in a suitably appropriate way. In that sense, our mistaken belief that they come from another star is no less silly than Ezekiel's vision of a godly messenger was several thousand years ago.

In truth they are from nowhere else but right here – appearing in great profusion because they come from all of the future to visit us and manifesting at any point throughout what to them is just history. They may differ in design because they come from various future eras. Yet again this accounts for what we see far better than does the idea of countless visitors from outer space paying scientifically absurd interest in earth.

Just as Hynek said, if UFOs travel through space they should appear at point A (let us say London), then later at point B (e.g. mid-Atlantic) and finally at point C (perhaps in Boston) after that. But there are almost no UFO reports witnessed by multiple groups of independent people at different locations that do not turn out upon study to have conventional solutions (usually as a meteor or space debris re-entry). One of the most significant features of the real UFO is its remarkable localization. It is also rarely witnessed arriving and departing the scene of its manifestation.

But what if we are looking in the wrong place to find these other sightings – i.e. we should not be scanning space but time? Say point A is not London and point C Boston but instead these should be different dates as the UFO travels back and forth from and to its year of origin? We may well have evidence that this does occur. Look at the following cases.

On 24 May 1949 on the Rogue River in Oregon, USA, five witnesses fishing on a boat observed a very distinctive object. A detailed investigation by the US military proceeded and that report was kept secret for many years. The witnesses made sketches of what they saw in full view in clear sunlight.

In essence it was an upper and lower hull separated by a row of convex apertures like ridges. There was a tapering tail with a patch on it and an overall smooth and rounded curve to the body. When first seen the object was standing on its edge and generating huge outpourings of light energy. It then just vanished suddenly.

Long before this report was made public Bill Dillon was one of many children and one incredulous teacher on lunch break at Ramridge Junior School in Luton, Bedfordshire, England. The date was during May/June 1957. They saw an object approach and move in an arc across the sky, almost spiralling downward and drifting across their path. Its motion was very similar to that at Rogue River.

In appearance the object was identical – not just similar, but absolutely identical. Given its unusual features that is very significant. The sketches eight years and 8000 km (5000 miles) apart are unmistakably of the same device. Several minor details cross from one sighting to the next. When I first saw this comparison, getting the Rogue River report from its release under the US Freedom of Information laws introduced in 1976 and comparing it with the sighting described by Bill Dillon, I was stunned.

Indeed, Bill had even told how at Luton the object had arched upwards and climbed almost vertically, then suddenly glowed brilliant white (as if its material form were turning into light energy) and vanished on the spot.

I have searched the records and found several other cases that match very closely. This includes one seen on the ground at Fort Riley, Kansas in November 1964 – where a close observation was possible for some minutes and, most amazing of all, a series of photographs taken at Mount Clemens, Michigan, in the mid-afternoon of 9 January 1967. Here Grant and Dan Jaroslaw took four Polaroid pictures of what to all intents and purposes seems to be the same craft. It vanished in exactly the same spatial direction (south-east) as did both the Rogue River and Luton objects. The case was considered genuine upon investigation by Dr Hynek and the US Air Force.

It has to be added that the Jaroslaw brothers wrote in 1976 and confessed to Hynek that, although they had shunned publicity at the time, the UFO was just a hoax concocted using a model. Perhaps so. But it was remarkable that they duplicated this very unusual UFO type so extraordinarily well – especially as in 1967 they could not possibly have known about either the 1949 or 1957 incident.

Even cases with human-like entities tied in show this sort of pattern.

On 21 October 1954 Mrs Jessie Roestenberg and her two children, aged six and eight, heard a hissing noise and went outside to see what they assumed was a new jet plane. Instead there was a flat-based object made of silver grey metal with a curved top and a large, transparent, dome-like window surrounding it. This was hovering at rooftop height looking over their isolated farmhouse near Ranton, Staffordshire.

Inside the dome were two figures gazing out of the window at the scene below as if they were tourists taking in an awesome but slightly wistful sight. They looked at Jessie almost compassionately. They were very human in appearance with pale skins, long fair hair and rather pointed chins. They both wore blue jump- or ski-suits, which looked rather odd for the time in question.

Mrs Roestenberg told me when I interviewed her 35 years after she first reported this sighting that as she watched the figures 'it felt like hours passed, but it must have been seconds. Time was suspended.' She also felt a tingling sensation on her skin 'as if charged by some kind of energy'.

As any mother would have been, she feared for her children and took them inside to safety. They hid under the table. Through the window the object was seen to arc around the sky then flash with a brilliant purple (seemingly ultra-violet) explosion of light and vanish into thin air.

This has become a celebrated case of alien contact – the first in Britain to be taken seriously. But was it something more exciting than that? One quarter of a century later, in March 1979, a woman talking to a man beside some shops in Sheffield was to relive Jessie Roestenberg's experience so closely that it seems hard not to perceive it as the same UFO.

They saw an identical disc with the large wrap-around window through which two figures were staring wistfully down. One was watching both witnesses. The other had hands behind his back as if more disinterested. It was a bit like someone who had seen it all before alongside a person experiencing this sight for the first time.

The figures wore blue ski-suits and were human-like with blond hair over their shoulders. Throughout the encounter an unusual stillness and silence descended over the area. It was 'as if they became isolated from time'.

The pattern can be traced through other cases. For instance, Mrs Ethel Field of Parkstone, Dorset, had the following encounter when taking in her washing in September 1977.

She heard a humming sound, looked up and saw a flat-based domed object with a large round window dominating the front. A huge explosion of light, a bluish-yellow colour, was pouring from the base. A peculiar patterned effect in the air was generated in waves beneath the object (it seems as if space was being disturbed by its presence).

Standing by the window were two human-like figures side by side. They seemed quite ordinary but had silver-grey one-piece suits on. Their faces looked rather thin. One of the men had his hands out of sight behind the base of the window, as if operating instruments and seemed detached from what was happening. The other man seemed to gesture downwards, pointing at Mrs Field, clearly noticing her. She panicked at this and fled.

Interestingly, she had held one hand up, palm outwards, shielding her eyes from the bright glare. This felt warm and tingling. For some weeks afterwards she had an irritating red skin rash which covered just that exposed part of the hand and, one seems safe to assume, may well have resulted from exposure to some mild form of electromagnetic wave energy emitted from the hovering device.

Another encounter comes from Mrs Sage of Medway, Kent, who was walking to the shops on the afternoon of 4 August 1980 when, unlike the other witnesses, she was fortunate enough to see the object appear. What she reports sounds absolutely nothing like the arrival of the Starship *Enterprise*.

As she described it to me, at first there was a small cloud, like a smoke ring, that formed in the air above a garden. Electric sparks shot from its edge and she jumped back instinctively from this. Then, 'this thing seemed to appear from nowhere in the place where I had seen the smoke. . .'

The object was small and curved, almost like a helicopter without tail or rotor blades. A huge transparent window dominated the front as it tilted down towards her and hovered very close by. She could see every detail of the two men inside. They were normal and human in appearance but with shoulder-length hair, one corn-coloured, the other slightly darker. Their skins were of a slightly unusual hue and their eyes blue. They wore one-piece ski-suits of a pale greyish colour.

As the object hovered only a few feet away the noise of traffic nearby became muted. Mrs Sage told me it was 'as if I was hearing it from inside something'. The strange mist surrounded the base of the object and clung to a picket fence separating her from it. It smelt sickly sweet.

Stranger still, Mrs Sage says she heard the two beings as they held a conversation just a few feet from her whilst still inside their craft. They were looking at her whilst doing so, both concerned and yet uninvolved. One remarked, 'You said it would be all right', then added, relieved, 'It's all right. She thinks we're army'. The other, rather sterner looking man, complained, 'All that work wasted.' But his colleague responded philosophically, 'Not necessarily.'

Mrs Sage did not see the object disappear.

The visual evidence for UFOs is much more limited than might be expected. Almost all the photographs that appear in books and magazines are dubious, in the sense that some are outright fakes and most are no more than misperceptions frozen into a UFO-like pose by the instant that the camera shutter was open. However, of the handful of attestable cases that have passed the critical examination, using techniques of computer enhancement, for example, almost all show features consistent with the pattern just outlined.

One of the most well-thought-of series of pictures was taken by a farming couple at McMinnville, Oregon, in May 1950. They made no money and sought no publicity from the two photographs captured on black and white film.

Measurements of the object in their pictures strongly suggest a domed craft with a flat base and turret on top, that was perhaps about 20 metres (66 feet) in size. We had nothing flying in 1950 which matched this object. We do not even have anything today. Yet its appearance is akin to those seen by witnesses like Jessie Roestenberg and has the features of those reported at both Ramridge and Rogue River.

Indeed Rogue River is only 240 km (150 miles) south of the position where these two pictures were taken and one of the other significant cases (the Williamette Pass photo described below) also comes from this same rugged, isolated mountainous area roughly midway between these two sites. It is rocky zones full of geological and tectonic activity where UFO sightings are rife. We must suspect that this is somehow important.

As usual, the McMinnville object appeared suddenly, disappeared just as rapidly and was seen by nobody at any point flying in between.

There are also cases where these objects appear to be associated with smoke or vapour clouds that are emitted by the phenomenon when it appears.

A postman at Namur in Belgium took a series of three images of a flat-based domed object. After remaining stationary it emitted a trail of smoke or vapour, which surrounded it like a shroud. It then shot towards the upper levels of the atmosphere, trailing a fine mist of particles in its wake.

On 16 January 1958 the scientific personnel on a Brazilian naval vessel were setting up a meteorological station on the uninhabited rock – Trindade Island – several hundred kilometres out in the Atlantic. This was part of the global scientific survey, the International Geophysical Year (IGY). Several members of the team observed an object resembling the planet Saturn that was surrounded in a fine mist or fuzzy vapour. This flew across the sea, circled the island and disappeared, but not before the ship's photographer got four pictures of the thing. At the time when the object passed by the ship the electrically operated winch failed to function.

As with the Namur case, computer evaluation found features difficult to hoax and enlargements of the Trindade Island object enhance its fuzzy – half here and half not here – impression which may be important. It even allowed one sceptic to propose the remarkable (but I think absurd) theory that this is a previously unfilmed daytime mirage of the planet Jupiter!

These photographs seem hard to relate to metal spaceships from a solar system beyond our own. They are more akin to a device that is distorting time and space by creating clouds of energy and phasing into and out of our perception as it moves.

There is one case which is even more telling. The date was 22 November 1966 and the location was 1520 metres (5000 feet) up on the Diamond Peak route across the Williamette Pass in Oregon. This is a remote, rugged and snowbound place to be on an early winter's day.

The witness and photographer was a college professor with a PhD in biochemistry. Not your typical UFO faker. He was atop the pass to film a granite outcrop but the slightly misty conditions made this difficult so he waited for the best weather conditions to arrive after two fairly mediocre attempts. His wait was rewarded by the appearance of an object climbing upwards at great speed from the tree-covered snow field directly below. The witness snapped a single picture almost by instinct but the object was gone before he could look for it again. It had simply disappeared.

The resulting image was later studied in detail by photographic expert Adrian Vance for *Photographic Magazine* as an 'exercise' to measure 'distance, speed and size' without himself forming any conclusions about the nature or reality of UFOs, although his report indicates his positive views. The UFO on the photograph was trailing a vapour beneath it, possibly a plume of powdered or vaporized snow being sucked with it by upward momentum. However, by far the most unusual aspect of this case is that there is not just a single image of a flattened disc on the picture. Although there was only one object – that is what the scientist's eye registered – the superior resolution of the camera, freezing a fraction of a second of time on film, had recorded three – stacked up like tiers of a cake.

Yet, stranger still, the object did not have the same dimensions in each image. This implied that it either decreased and increased in size or moved away from and towards the camera during the very short period that the camera shutter was open. The motion, if it were that, was like a plate when dropped into a bowl of water, sliding from side to side almost like a pendulum bob. The same effect is noticed in the autumn when leaves fall from trees as they flip flop down through the breeze. Indeed, this same effect is a feature noticed often enough in UFO cases that it has even been given the name – the falling leaf effect. But nobody had filmed it in operation.

Yet there is another factor all the more significant. If this is a single object (as it was seen to be) rising into the air and either changing size or swinging to and from the camera like a falling leaf to create an illusion of doing so, then the camera ought to have recorded a continuous blur of motion. It does not do so. It shows three discrete images of the object at three separate spatial locations.

Assuming that the photograph is genuine, and until very recently nobody to my knowledge has attempted to establish any reason why we should not assume so, then this UFO seems to have inexplicably phased in and out of our perceptual reality three times in such a short fraction of time that the eye was deceived and saw this process as a continuous motion. This is just like the way in which we watch a movie film. That is an illusion created by showing a series of slightly changing still images displayed before the eyes so fast that they cannot record them as separate images and must see them as a continuous motion.

In the Williamette Pass case the camera appears to have seen something the eye could not – and recorded how this object was literally fading in and out of our space-time reality. Moments later it vanished altogether. To the eyes of the witness – as indeed to the eyes of many other witnesses to similar phenomena – it just seemed to vanish on the spot. But if this photograph is to be trusted, we can see the mechanics of how that happened. By far the best explanation for this behaviour is that this object was travelling through a space-time field rather than simply upward through space like a rocket ship or aircraft.

However, as we go to press, a case which has remained dormant for many years was re-opened through remarkable research from Californian analyst Irwin Wieder. His paper, 'The Williamette Pass, Oregon, photo revisited', appeared in the *Journal of Scientific Exploration*, Summer 1993*, and presents an extraordinary attempt to demystify the image as a roadside sign post. He admits that he came to this conclusion only after a decade convinced that this was a very important photograph and was persuaded otherwise simply by the wealth of evidence he uncovered.

* *Journal of Scientific Exploration*, ERL 306, Stanford University, Stanford CA, 94305-4055, USA.

Despite the seeming absurdity of his suggestion that this complex triple image could be a road sign, Wieder provides excellent grounds for taking that theory very seriously indeed. He includes comparative photographs showing images of considerable similarity to the Williamette Pass photograph that he took of both mock signs made to his own specifications and another on the same roadway in question. His evidence looks strong enough not to be rejected out of hand.

However, if Wieder is correct then a witness of impeccable scientific background, by whom he admits to having been impressed, and who has never made money or sought publicity from the photograph, would, at the least, have had to confabulate his story to alter crucial facts. If the picture is a road sign (and no sign is present at the location today, although there is evidence one might have been there in 1966) then the scientist could not have been standing at the position where his other photographs of the Peak were taken that day or have seen the object rise into the sky as his testimony alleged. He must have taken the vital picture from a fast-moving car at a different spot on the highway, causing the adjacent sign to blur but the background scenery to remain in focus. He has denied doing this.

In 1980 and 1981 there were two fascinating dramatizations of the time-travel theory concept in BBC TV movies, now available on video. They were entitled *The Flipside of Dominic Hyde* and *Another Flip for Dominic*. The time-travel stories are themselves well worked out and excellent entertainment, but they are set against the backdrop of historians from the future routinely overflying the twentieth century for data-gathering purposes. It is known from future archives that these flights were often observed by our technology and the broad population density that exists today but that the myth of alien visitors was rapidly built around them. The future traveller is seen observing one such BBC documentary on UFOs, which ironically I was involved with, and marvelling at the stupidity of our preconceptions.

However, whilst contact is strictly prohibited between time travellers and people in the past to avoid the fear of the dreaded paradox rearing its head again, there is an understated awareness amongst future historians like Dominic Hyde of the need to cover tracks where necessary. This can be done by playing up the image of visitors from space should the need ever arise. No doubt in other eras they similarly would have emphasized whatever view was current amongst the population – be it as gods, messengers from heaven, wizards or witchcraft.

As such, we can imagine how the cultural belief in UFOs as alien spacecraft might even self-perpetuate itself by a sort of deliberate reinforcement by the actual agents of the phenomenon themselves. This would serve to deflect attention away from the truth about who and what they are.

I suspect that this is exactly what would occur in any situation where travel from the future to our present day became quite commonplace. Probably it would be regarded as unwise to allow the concept of time travel to become accepted too soon. A self-perpetuating mythology is the perfect answer, with an engineered smoke-screen creating the rather dotty idea of alien visitors. What scientist or government would take that too seriously? As such, these forces would pose little threat to regular visitors from tomorrow. Otherwise, in our heavily defenced skies and with our post-World War 2 militaristic frame of mind, time travelling would be a risky pursuit.

The concept of UFOs as time travellers from the future is oddly satisfying in a theoretical way. It makes sudden sense of many features of the UFO evidence that otherwise seem difficult to reconcile. For instance, why the 'aliens' are so human, not only in appearance but in their ability to speak our language and, even more significantly, in their thinking processes and emotions. These rarely display even a hint of anything extraterrestrial and seem so obviously human that they can be nothing else.

Equally, we have the problem of 'cultural tracking', where the UFOs rarely, if ever, display evidence of a technology or knowledge beyond our own. That would presumably be something that was carefully hidden to avoid a time paradox whereas aliens would have no need to do the same thing.

There is much else, e.g. the long historical record and the way in which a series of different UFO types crop up on an unchanging basis throughout history, showing no signs of progressive development as we would expect if these are visitors who have shared that time with us. Just look at how our own technology has improved in 2000 years. The reason why the same sort of UFOs were seen way back then as are now being seen today is rather hard to explain as a growing, thriving alien race who have had thousands of years to boost their own technology just as we have done.

Of course, if UFOs are from, say, a dozen different eras in our future then each one could visit any part of our past. As a result we would see a dozen different basic UFO types scattered liberally throughout history with no trace of any scientific progression. A UFO from the year 10,000 could visit 50 AD just as readily as one from the year 2010 could do.

There are many little pointers like this which indicate the way in which the UFOs-as-time-travellers theory transforms previously puzzling data. And, as we have also seen, it helps to explain internal features of the UFO evidence itself – such as the behaviour of these craft, the raw details of the few genuine sightings of strange machines and, equally, the rather ludicrous planets of origin and silly extraterrestrial data that a witness is offered when contact with the entities takes place.

The chat-up lines of the 'alien' seem as if they have been invented from reading comic book or science-fiction novels. They sound like stringing along the witness with tongue-in-cheek remarks (e.g. claiming that they come from a planet hiding out of sight behind the sun, or a 'small galaxy near Neptune').

In fact, if they are time travellers masquerading as alien visitors, leg-pulling may well be exactly what they are doing – in fact, deliberately misleading with undoubted relish. One can imagine time travellers returning home to swap tales around the fire about the dumb yokel from 1994 to whom they have just told a wild story of their alien origins. As historical archives will then show this tall tale was accepted as 'true' by tabloid newspapers, TV chat shows and best-selling books penned by UFO buffs!

However, there are other ways to look for evidence of time travellers in our past. Are there any tangible signs that they have left their mark on society across the millenia?

I could spend some time delving into the work of writers such as Eric von Däniken who made millions of dollars in the 1960s by suggesting that aliens had visited the earth long ago and created a technology whose remains can be seen today. He cited what look like batteries found in the Middle East and dated long before electricity was discovered and countless other memorabilia in his string of books.

I will not delve into these, partly because some of the ideas have been disputed by later work and partly because it would be needless duplication. I think it has to be remembered that human achievements should never be underestimated. There is no reason why we should not have built pyramids or erected Stonehenge just because it seems incredibly difficult to do. It is perfectly possible that some brilliant inventor would produce something so far ahead of its time that it remained a curiosity until science caught up.

None of this proves (or disproves) that any ancient wonder was created by aliens, although some of the data is certainly interesting. However, it should always be remembered that in von Däniken's theory 'ancient astronaut from another world' can always be substituted with 'time travelling visitor from our own future'. In that sense, some of his research might be worth interpreting in a whole new light.

Equally, one can imagine some rogue time travellers from the future possibly attempting to guide humanity along a better route. Were Leonardo da Vinci's incredible inventions (of submarines and flying machines, which seemingly could have worked with access to technology to match his vision) simply the result of a medieval genius far head of his time, or were they given to him by someone from a future age when trying to change history?

Most religions of the world have at their root powerful visionary experiences with human-like figures with magical powers and craft who offer access to advanced knowledge and provide moral and ethical codes to accompany them. One can imagine how some biblical stories or, more recently, the formation of the Mormon Church, may at least in part be the consequence of carefully staged visitations from the future.

Of course, saying such things will come over as reprehensible and offensive to many people with strong views, and I am not for one moment saying that this is what happened. I am merely conducting an academic exercise and cautioning why this sort of evidence will not suffice.

However, there are more concrete things that we can look at which offer tantalizing glimpses of a possible reality.

Firstly, there is the puzzle of the ancient cave paintings found in paleolithic (or stone age) sites around Europe. These date from a time long before civilized man settled southern France and northern Spain and range from between 12,000 and 32,000 years in age.

Many of the pictures etched onto the walls of caverns at sites such as Lascaux and Altamira are of animals such as bison or long extinct mammoth. But there are some truly extraordinary images as well. There are literally dozens of drawings of what for all the world appear to be UFOs exactly like that filmed at Williamette Pass. These are shown on the ground and in flight. One of the most amazing, at Niaux in the French Pyrenees, displays two objects side by side, one apparently giving off the same sort of trail of smoke or sparks reported in so many of the cases just discussed. This picture is estimated as being 14,000 years old.

If these were isolated examples one might well be dismissive of them. But they are not. They are all over the cave systems, as much a part of the imagery as are the animals that we know were really there to be carefully illustrated by these very first portrait painters. So what are these strange objects? Were alien spacecraft visitors to this part of Europe all that time ago – or are the ancient artists depicting some of the earlier voyages of the same time travellers whom we still see today?

Geology also has several mysteries to confront us with. One is the riddle of entombed animals and out-of-place fossils.

Fossils occur when an animal dies and is submerged, e.g. by sediment at the bottom of a river or in a mud pool. Not only does this preserve the animal from predators but over the centuries as the deposits harden to form rock it traps the bones. Either these dissolve and a cavity forms that is, in effect, a cast of the creature's shape, or the bones are replaced by rock itself through a process of chemical transformation. Tens or hundreds of millions of years later that rock, which might then have been raised from the sea floor to the top of a mountain through geological processes, could be excavated. From the midst of this solid rock we would find the bones of an animal that died so far back into history it is almost unimaginable.

Indeed fossils are vital clues that help us date the processes by which the earth has evolved. By charting the progress of animal types we can date the rock in which they are found. Obviously, this rock strata must be more or less contemporary with the death of the animal concerned.

The problem comes with cases where a fossil appears in rock which clearly dates from a time period far removed from that when the discovered animal was living. As the study of fossils grew this was sometimes because we had an uncertain knowledge about geological dates. But there are still cases where this seems not to be the answer – fossils of animals in rock way out of time with what life forms should exist within that era.

Few theories make much sense, but one is that the animal in question somehow travelled through time – either forwards or back – and died in the wrong era. It then by chance became fossilized to leave a permanent riddle that we still struggle to comprehend countless centuries after its passing.

There are reliable stories of frogs, for example, discovered within lumps of coal excavated from deep mines. The coal formed under intense pressure perhaps 300 million years ago. The animal fossilized within did not first live until more than 250 million years later than that!

176

As yet there have been no discoveries of fossilized human beings in very old rock. If a time traveller from the future died in the very distant past and was submerged in mud perhaps that person could theoretically survive as a fossil. The earliest known human-like beings date from around 35 million years ago – extremely recent in geological terms. This frontier is most unlikely to be pushed too much further back. If recognizably human bones appeared in, let us say, the coal beds from the Carboniferous geological era almost 300 million years ago then that would be as close to scientifically impossible as one could get. The only explanations would be alien visitors that just happened to resemble ourselves or, rather more plausibly, time travelling humans from the far future.

In fact, we have something almost as good. Not a human being exactly, but apparent evidence that human beings were around far back in history, even further back than when dinosaurs began to dominate the earth. This evidence comes in two forms – traces of human activity and artefacts.

There are a number of cases where animal tracks have been found in solid rock which long ago was wet sand that solidified over the millenia. These range from worm burrows to footprints of giant dinosaurs. Sandstone also often displays dunes and other evidence of the beaches from which it may have formed long before mankind walked the earth.

Indeed, anyone arriving by train at Liverpool's Lime Street station, for example, is treated to a fine view of a cross-section through some fossilized dunes etched into sandstone rock from which the tunnel was excavated during the last century. The dunes themselves once stood in a desert upon which dinosaurs almost certainly walked but upon which no human feet could possibly have stepped – unless they had access to a time machine capable of traversing hundreds of millions of years to get them there.

Yet in similar locations around the world what do appear to be human footprints have indeed been found! You can see them in various parts of the world. In the Mississippi Valley of the USA a block of limestone quarried for building in 1822 was found to contain a perfect set of prints of unmistakably human feet. Toes and heel markings were readily visible.

In Nicaragua several tracks were found well below the surface on volcanic pumice covered by harder rock and several other volcanic layers totalling several feet in thickness, thus indicating their antiquity.

At Carson, Nevada, in 1882, during rock excavations, one area was found in sandstone where many animals had evidently walked millions of years before. They had left prints and amongst the animals were marks left by a human being. They were, however, very large – about 45 centimetres (18 inches) in length.

Several other well-attested cases exist. Science usually either ignores this perplexing evidence or seeks a way out. The most popular rationalist interpretation is that the prints are really not human after all but a large mammal such as a giant sloth (used to explain the Carson, Nevada, prints on the basis that these creatures did coexist with the other animals whose prints are found there).

The other much-used solution is that ancient peoples etched them into rock for ceremonial purposes. This idea can only rarely be made to work – that is when the rock is exposed at the surface. If, as has occurred several times, it is excavated from deep underground that is, of course, out of the question.

In the Nicaraguan example, despite the depth of the rock, the nature of the terrain, volcanic rock, suggests that these prints may not be old enough to be incongruous. Volcanic deposits can grow very quickly in geological terms because huge volumes of ash may be ejected by a single eruption. Any area active for only a few thousand years could easily cover footprints by many inches and rapidly turn into rock.

Some classic prints found this century at the Paluxy River in Glen Rose, Texas, are especially impressive, because they show what seem to be human beings walking side by side with dinosaurs – indeed at one point an overlain human foot crushes what must have been a recently formed dinosaur impression by walking all over it! This is, of course, something we know cannot have happened (at least, not without time travel, that is).

The dinosaur prints are undeniable. The seemingly human ones in the same rock more than 100 million years old have been a cause of concern to scientists for some time. But computer simulation work in 1986 suggested that their human appearance might be an illusion.

Several researchers, using this data, have now concluded that an unidentified type of dinosaur made them by walking on its heels (as opposed to the normal toe-driven view of dinosaur walking skills). The raised toes had not formed proper fossil markings but were filled in by mineral streaks, furthering the impression of a human appearance. The matter is still somewhat controversial.

However, these solutions do not entirely clear up the problem of these fossil footprints, even if we accept all the assorted explanations as valid. There remain some quite baffling examples.

In 1927 at Fisher Canyon in Nevada limestone laid down almost

200 million years ago contains an imprint of a shoe with a leather sole that seems unlikely to be anything but that – even if this is quite impossible by any normal definition.

Yet by far the most extraordinary is a cast of what looks to be a shoe 38 centimetres (11 inches) long that was uncovered from the middle of a piece of limestone in Utah. The rock dates back 500 million years and, indeed a squashed trilobite is found beneath the foot, seemingly crushed by the person who walked over it. Trilobites were extinct long before even the most primitive ancestors of mankind first appeared. If proven, this print can only come from a time traveller who went 500,000 millenia into the past!

Also very difficult to account for are the artefacts found embedded into rock. These include some remarkable items often found by accident when a lump of coal was dropped before being put onto a fire.

The coal deposits have come from hundreds of feet down in a mine and were laid down hundreds of millions of years ago. Yet inside these deposits have reportedly appeared a gold chain complete with cast inside the rock, found by a woman in Illinois, USA, and several metal nails, including one cut from inside a lump of limestone from Scotland. This was laid down within a shallow sea countless millions of years before there were any human beings to sail any possible ships from which it might have been dropped. And quite how a coin dated 1397 was found inside some 300 million-year-old coal is very difficult to imagine without time travel. But such a find was made in 1901 and published in *Strand* magazine.

All sorts of strange ideas have been proposed to account for these mysteries – ranging from teleportation through space and time to Robert Rickard and John Michell's weird concept of the 'womb' of the earth 'growing' these items within itself through some inexplicable process.

Frankly, I think that much the most plausible solution is also the most obvious. That these are simply examples of objects displaced through time as a consequence of time travel when it does become possible. By pure fluke these items (from many footprints made or artefacts dropped) were embedded in rock or fossilized along with creatures native to the time in question.

If time travel were ridiculous or unsupported by any other evidence this would, of course, be an extreme concept. But the evidence for this ability, whilst not overwhelming or absolute as proof, is at least strongly suggestive that this may well be what is happening (or about to happen!).

I think that one day we will travel back in time. When that occurs our descendants will be able to return – not just to 1994 or 1397 but to the year 500 million and 94 BC! As a consequence they will inevitably sometimes leave evidence of those visits which, although in the one sense they result from trips that have not yet happened, in quite another very real sense did happen – long, long ago!

Conclusions

Do It Yourself Time Travel

IN THIS FINAL CHAPTER I will summarize discoveries that we have made about the amazing nature of time. In doing so I will present a series of practical experiments that can be carried out to help anyone to explore the deeper levels of the temporal universe.

Some of these exercises can be conducted by anybody who is reading this book, working alone and with nothing much more than the perseverance to try them and then see what happens. Whether you believe they will work or not, what have you got to lose? The potential rewards are as fantastic as any scientific breakthrough that Albert Einstein might have made. The repercussions, for our view of how the universe works, are truly awesome.

The other experiments that follow probably require a broader canvas – needing the co-operation of groups of people, perhaps even input from the media. None requires vast sums of money to be attempted. I am hoping that many of them will be set in motion to coincide with publication of this book. Who knows what we might achieve by such a concerted effort?

I believe that three important things have been established by the evidence we have surveyed within this book.

Firstly, that time travel in a variety of spontaneously occurring ways clearly does take place. Sometimes this is through our mind or consciousness. As we all have consciousness, that particular route towards mental time travel is readily available to everyone who chooses it.

However, there also appear to be natural energy forces of great strength that seem capable of creating rifts in space and time. The series of consistent stories involving motor vehicles that encounter

181

strange, drifting clouds point firmly towards how these might operate in the world about us.

We cannot study these to order in a laboratory as they are at least as rare as ball lightning seems to be. Science still struggles to comprehend that. But they exist. Now we also know that these clouds of energy are out there and can stop deifying them as alien kidnappers or mystic forces. As with all things, from electricity to nuclear radiation, understanding them as part of our real world is the first step towards taming their raw power.

Secondly, we have seen that science is moving inexorably towards that day when time-travel devices will be built. Indeed, some say, the first prototypes already exist. Whether practical time travel will first occur by way of TV images, flying machines or controlled altered states of consciousness is open to dispute. But the hunt is on and the day when possibility turns into reality is clearly not too far away. At the very least, we now know that time travel is no longer science fantasy.

Finally, given that time travel in some form is almost certain to come about – probably sooner, rather than later – this implies that we ought to have evidence of that coming reality right here and now. Unless there is some law of time travel that prevents it taking place, then voyages from the days when time is traversable back into our present and our past should already be a matter of record relative to our perspective. Evidence of these travellers may well be out there waiting to be discovered.

We simply need to have the courage to look for the seemingly incredible amidst the mundane trivia of life. We found broad hints that seem to suggest how time travellers may have been around in earlier days. A concerted effort to look for more hard evidence may provide that one, conclusive piece of data that proves this amazing truth for all to see.

These conclusions are astonishing enough. But I do not think that the facts would suggest them to be ridiculous. Rather those facts support, even if they can never prove, that what I propose seems to be the truth. However, the real excitement comes from knowing that time travel is not just the province of quantum physicists with billion-dollar budgets to play around with. It is potentially available to every person on earth.

We simply need to know where to look – so, let the journey begin.

EXPLORING THE PAST

1   Do you remember the day you were born?

A simple way to explore the idea that mind can span time and space is to question very young children. Ideally they should be no older than seven or eight years, preferably between three and five. They need to be able to convey images to some extent through use of language and to be relatively free from social contamination that comes by watching much TV, reading and school education.

If you have your own children, so much the better. This must be a parental decision. Never ask questions such as these of children who are not your own without checking with their family first. It may offend a moral code or be considered spooky by some adults. Always respect their wishes.

Assuming there are children that you can talk to, then tape record their responses. Treat it as a game with them. Do not appear serious or tell them it is a scientific experiment. If they want to sketch or paint scenes then let them. They may feel better able to express themselves that way.

First ask them this question: *Do you remember the day that you were born?* Never push them. If they say no, accept that. You could try enquiring about the first memory that they have, but young children are not as time indoctrinated as we are and they may not have a sequential set of memories in quite the way that you would expect them to do.

If you do get any response and it is clear that the children are enjoying this and are in no way concerned by the questioning, then next ask: *Do you remember any time when you were older than you are right now – perhaps, it may have been long ago and you were even somebody else then?*

Do not elaborate on this. To say more would be to lead the child. This set of questions already does lead them quite enough.

Record whatever they say – if they say anything of merit – and smile and laugh with them, if they want you to do. But never push the experiment beyond five or ten minutes. After that the child will want to do something else.

Of course, if anything emerges it could well be just imagination. But keep a record of any such 'memories'. You might even try to check them out!

2   Historical anachronisms

It can be fun to seek out historical records somehow out of time, but it can also be a tiring and seemingly unending process. It is recommended to anyone who likes to do hours of research scanning the microfilm records available at most major reference libraries. You can go there and use them at your leisure to access old newspapers, both local and national, sometimes dating back well into the nineteenth century. Aside from the fact that this is an interesting exercise in its own right, to be steeped in the history, culture, even the adverts of the time in question, there are other potentially fantastic rewards available.

What you are seeking is an anachronism – that is, something which appears instantly curious and out of step with the day in question. It is unlikely to be as obvious as the example I cited earlier in the book – a coin from the future bearing an unborn monarch's head found in a till by a shopkeeper. Instead it might just be a seemingly amazing prediction so accurate that it could, perhaps, be based on direct future knowledge. Or a person might behave or talk oddly, seeming to be aware of future technology and telling people of how the world will be many years hence. Frankly, you cannot know up front what you are looking for in this sort of quest. But if you find it then it will leap out at you from the page.

Another task you can attempt, if you get bored with this open-ended search, is to examine press photographs from the scenes of major historical tragedies or other great events. You can draw up a list of dates and events to check before you visit the library – the fiery doom of the *Hindenburg* airship, the crowd of rescuers at Aberfan – the choice is yours.

This experiment is more rewarding to do but simpler if you can take hard copies of the newspaper pages home. Then you can study the faces in the crowd at leisure with a magnifier. You are looking for one person turning up at several different events scattered widely through time and space. Someone who might be a voyeuristic time traveller. Such evidence may not exist at all or may be very hard to find even if it does. But how will we know unless we try?

This sort of experiment is not easy. You need a stern resilience to keep at it. It is not unlike the fossil hunters who spend months or years before they come across the 'big one'. As a tip, start by looking at the newspapers around your own birth date. This adds extra spice to the search.

3 Is that a UFO I see before me?

Another experiment to try can be done in association with your search for anomalous people in the crowd at the scene of major historical events. Scour the records for UFO cases that may correlate with the event. By this I mean a strange object in the sky that was seen in the same geographical location and on at least the same day as some big event was taking place.

It is quite possible that this will not be mentioned in conjunction with any news accounts of the historical incident, because nobody at the time will have considered there to be any possible connection. So you will need to examine other parts of the paper and – keeping the date of the event in mind – the letters columns in subsequent days and weeks where anything 'quirky' such as this may well have been reported by a reader.

Remember that prior to about 1955 the word UFO would not be used and that earlier than 1947 not even flying saucer will serve as a hook to drag you in. Earlier records, if they made the press, will have been talked about in many different, often cynical, ways – and the skill is to decode them.

You may read of strange lights, ghostly shapes, unusual meteors, peculiar comets, unknown aircraft or (particularly between 1890 and 1920) airships that appeared out of nowhere. There will be other descriptions too, possibly including things like religious visions or signs from God.

Contact UFO organizations also (see reference section). Some have computer archives for post-World War 2 cases and could correlate these.

Keep a score card of each historical event that you scan and whether or not anything turns up around it. Any object reported some miles away or on the day before or after should go down as a 'half hit'. Anything else is a miss. A 'full hit' occurs only if the aerial phenomenon was reported within a few miles of the scene of the event and on the same day.

As something like 15,000 UFOs are reported worldwide each year (of which most are not real UFOs at all, of course) that means about 50 are occurring somewhere on earth every single day. That will help you keep track of the statistical odds should you start to turn up meaningful results.

Of course, in theory we might expect time travellers to visit major events in the past for historical observational purposes. If they were out there watching, someone may have seen them!

NO TIME LIKE THE PRESENT

1 The power of the mind

I tried this as a prototype experiment on a small scale a few years ago and it worked astonishingly well. The idea is to demonstrate that we have a mind that can access information without reference to space-time and to which, by all mundane definitions, it ought to have no right to gain access.

The experiment is simple and unlike most tests of parapsychology, which are dry and dull laboratory probes without any emotional component, this has all that. As we found when studying time anomalies emotion is often the key to them happening at all. So to ignore that fact is folly. Equally, by theoretically involving dozens, thousands or even millions of readers at once this test is virtually assured of producing some level of success.

Although this ESP-eriment (as we called it) will work best through a medium such as a national newspaper or magazine, it can be tried with just a few friends on a smaller scale at a party, although obviously (unless it is a real hot party!) without using celebrities for the guinea-pig.

In the prototype I had designed there were six photographs of famous people printed side by side and chosen to be of the same basic type (i.e. all currently popular on TV). Those participating knew only that one of those six would really take part in the experiment on a given date and time. The other pictures were there as psychic red herrings.

At that given time, for some minutes, the celebrity was to imagine him- or herself in a particularly vivid or evocative location and try to picture that firmly in their minds. Anyone who wanted to join in the ESP-eriment simply had to sit and relax in a quiet location (preferably speaking into a cassette recorder to avoid breaking their reverie as they would have to do to write down notes). They could do so either at the appointed time, or if they preferred by describing their dreams the night before the test. The participants spoke of anything they visualized, or more importantly, any emotions that they felt. These tend to get past the doorman more readily.

The experiment is three-fold. Firstly, we have the straight statistical choice. Which of the six celebrities do they 'feel' is really taking part? The odds can be easily calculated. An anticipated 16 or 17 per cent should get the correct answer by guesswork. In fact, 23 per cent did so – well above chance level.

The second part of the experiment was more difficult to score. Quite an array of images and feelings arrived but some were remarkably precise. In gauging 'mood' or 'evocation' of the setting the success ratio was high.

Thirdly, we had the inbuilt test where a participant was not trying to alter their state of consciousness to reach the celebrity. They were already in an altered state of consciousness – via a dream. Using these as the basis for the imagery should increase the success ratio. Indeed it did exactly that in my dry-run ESP-eriment with about 100 submissions.

Fully 70 per cent correctly got the sex of the 'celebrity' correct (chosen candidates split evenly between male and female). Remarkably 40 per cent identified the correct celebrity by this route – way above chance expectation.

What this seems to indicate is that in altered states, such as dreams, the mind can roam free and access information that it is harder (but not impossible) to pick up whilst in a wakened state. Equally that 'raw' data (such as general moods and the essence of being male or female) is easier to tune into than more specific data.

There is every reason to try this ESP-eriment again on a broader scale – indeed as many times as are practicable, because it is a simple, fun way to demonstrate that our mind is not as limited as we think it is.

2   TV or not TV – that is the question

A related experiment could be attempted via a live TV programme. It may test this in an even better way. Have ready a series of short video clips that convey a specific and sharp emotion (e.g. joy, sorrow, peace, anger). Those participating would be asked to survey their dreams the night before the programme goes to air during which one of these clips will be transmitted live. Which of the clips is screened would have been chosen by some random method the evening beforehand, i.e. just before these dreams, but its contents would be known only by the experimenter. Those taking part would have to decode from their dream images the night before transmission the essential emotion of the clip that they are about to see. They can then phone in to the show with their 'vote' on which clip will be aired.

What will happen? Only conducting this experiment will tell us that.

BACK FROM THE FUTURE

1   Warning Signs

A very simple research project that can be carried out is to maintain a record of all the anomalies that occur immediately prior to either an earthquake or volcanic eruption. For most of us it is unlikely one of these will happen on our doorstep all that often, so this requires a careful record of all media stories about such disasters.

That said, oddly enough, I have experienced two small tremors in England and none (so far) in the time I have spent on several trips near the San Andreas Fault in California. Prior to one of these British quakes I noticed very unusual behaviour by the birdlife, which seemed to be aware by some minutes of what was about to happen and stopped flying to enter an Oz factor stillness.

Scientists, particularly in countries such as China, know that animals seem especially attuned to the onset of such natural disasters but do not fully understand what this 'warning voice' may be. All too little research is being carried out. So there is definitely scope for you to seek out patterns and trace any anomalies that might help us work out by what process this wildlife precognition seems to follow.

2   Project 'Future Oracle'

The opportunity exists to create a new type of premonition bureau that seeks to collate dreams and design an interactive computer software programme to marshal the information. It would aim to issue probability warnings from a combination of data and suggest in advance what may occur. This will take time and needs assistance from a range of people. I would be keen to hear from anyone who belongs to one of these categories.

Firstly, computer programmers who think they could help develop the interactive software that could define probability ratings for coming events by utilizing all of the input data.

Secondly, media sources which might be interested in promoting such a venture in exchange for probability warnings and stories about success after the fact. We would need to keep the project in the public eye for a period of time long enough to maintain a steady flow of incoming data.

Thirdly, anyone who thinks that they have strong dreams or experiences that do see future events on a regular (or semi-regular) basis and who would be prepared to document them with this project. These would be on special forms that we would design for the purpose and would have to be completed as soon as they had a dream which they felt strongly about. We would have to confine submissions to non-personal events or trivia – i.e. to world dramas or those involving celebrities for the purposes of the experiment.

How long we need to get project 'Future Oracle' off the ground depends on the support received. Potentially, it may have great benefits.

As I write a Luton man who believes he often has visions of the future has claimed a startling hit that shows the potential. On the night of 16 July 1993 he awoke from a vivid emotive dream and recorded details right away as he has become trained to do. In this dream he had seen a 'fair or display', the figure 2, and a plane crashing into the ground in a loop. On 23 July he had another dream in which he saw 'fireworks' exploding in mid-air, rockets falling and two parachutes descending from the sky.

Chris Robinson felt so sure these meant something that he called his local crime squad. A police sergeant there confirmed to *Psychic News* that he had fielded a call from this man on the morning of Saturday 24 July with a prediction. Nothing was done with this information. What could be done?

However, only hours later, Chris Robinson heard on the TV about an airshow in Fairford, Gloucestershire, that was happening that very day. Feeling sure this was what his dreams were about, he drove 100 miles to get there, arriving at 3 pm after the gates had been shut. He can prove his attendance and claims he told a policeman on duty outside that he had to get in to the show, because two jets were going to collide. Moments later, in full view of both men, they did exactly that. Two Russian MIGs came too close in a manoeuvre and exploded in flames. Both pilots parachuted to safety and the aircraft erupted in flames and corkscrewed into the ground, miraculously causing no injuries to the crowd – as shown spectacularly on TV newscasts.

Think what a computer-based premonitions bureau could do if we had more information such as this coming in advance on a regular basis.

3   Join the chain gang

Something that anybody can do to try to seek out future events, even if they are as psychic as a rock, is to find chain reactions discussed on pages 120–37.

As we saw, these fall into several categories. They include personal coincidences happening within one's own life, mass trends – such as sudden changes in human behaviour patterns causing seemingly odd cultural events to occur – and those clusters of global incidents (some trivial and others not so minor, some obviously relevant and others more obscurely predictive) which all seem to emphasize and reinforce a particular trend or theme.

Whilst these look obvious when put together after the event, the hard part is to understand where they are leading us in the days or weeks prior to any event that they foreshadow. That seems to be easier said than done.

As this is an entirely new field, which nobody is researching so far as I know, there are no ground rules that I can offer you. Nobody has been looking for chain reactions before (except to some extent John Webb and me) and so we do not know much about how the process works. We will only find out what the mechanics are if you start to look for more such patterns.

I suspect that they occur on a large and small scale according to the importance of the event that they forewarn. Critical to the size of the chain reaction may well be the event's impact on our collective consciousness as a society. Therefore, we would expect a small chain reaction in a situation like the July 1993 Fairford airshow disaster, for instance (although one may have been set in motion that nobody detected). But there could be a bigger one if a President or Prime Minister were to die, simply because more people would be emotionally affected by that sort of tragedy.

Therefore, it is likely that at first we will be collecting chain reactions mostly after the facts that they relate to, analysing a situation that has occurred and seeing what patterns preceded it and might have been used to predict or prevent it. As nobody has done this with any major event in the world during the past, the field is wide open for anyone to start a retrospective search. There must be countless chain reactions that accompanied many of these mega-events. We just need to find them.

The aim has to be to figure out how to recognize the chain as it is still unfolding so that we can feed that data into Project 'Future Oracle'. Then, hopefully, we can do something positive with this very odd discovery.

4 The dream diary

Despite the fact that J.W. Dunne first wrote of this more than sixty years ago the dream diary is probably still the easiest way that anyone can try to experiment with time travel. But we have learnt things that we can use to good advantage to improve his methodology.

The principle is the same: to maintain a 'log' of your dreams over a period of time and record as much detail as possible from out of them, with special emphasis on dreams that seem vivid and have great emotional impact (whatever the emotion may be). These are the ones to look at as possible premonitions.

It is not necessary to do this experiment constantly – as that can become a strain. But you should always record any dream that seems powerful and if you have a 'gut feeling' about it. If so, make sure you let someone else know about it right away. Ideally this should be by submission to Project 'Future Oracle'. I will be happy to receive these dreams from you. But tell someone. Sceptics always claim that you only thought you dreamt something in advance, but if you have not shown your dream diary to anyone to prove this, then a few years down the track you may doubt that you really did dream of the future. Words in a notebook convey little after a while.

Probably the ideal is to run the dream diary experiment for a couple of weeks and repeat it every few months. This avoids stress. You will be surprised how easy it becomes to remember dreams after a few days practice. Many people think that they never dream or that, if they do, it is seldom and with few details. In fact we all dream for several hours every night. We just forget most of them as soon as we start moving about because dreams only enter our long-term memory store if we make a concerted effort to make them do so. Otherwise the doorman kicks them out pretty quickly.

In Dunne's day he had to have a notepad by the bed and write down what he could remember as soon as he awoke. That is how I started too – and it does work. But the act of opening your eyes, focusing on the page, fumbling about for a pen and writing all accelerate the process by which our dream memories dissolve, and it is far better to have a portable cassette recorder by the bed and speak into this. Keep your eyes closed as much as possible and your head still relaxing on the pillow.

You may feel awkward at first (especially if you are not alone in bed). But this is much the best way in the long run.

Do not check your dream tapes against reality every day – unless something in your mind urges you to do so because you think you have scored a hit. Do it about every three days. If you recall dreams in as much detail as you are likely to do after a while, this will be the maximum time period you can wait, otherwise you might spend hours listening to your half-awake self mumbling about things that will be largely nonsensical. By far the majority of your dreams about the future will hit within 72 hours in any case.

There are some tips that can improve the chances of scoring a hit.

One idea is to set your alarm for two hours before you are due to get up. If you can stand such a masochistic practice, the dreams that you have in the period after you switch off the alarm and fall back asleep seem from my experience to be about 50 per cent more likely to relate to the future. Nobody seems to know why that should be.

Another good suggestion, especially if you follow the above advice, is to start your dream diary experiment just as you are about to have some kind of new experience, such as a holiday, or perhaps a new job. The unexpected data input coming into your mind and, possibly, the increased need that the mind then has to work out future scenarios of behavioural responses through dreams (one of the key purposes of the dream in the first place, according to most psychologists) all seem to enhance the likelihood of some of those new experiences being foreseen by your mind.

Finally, since dreams of the future appear to describe your own emotional response to things that occur, the best way of having dreams about non-personal matters is to train yourself to watch a lot of TV news – especially in the morning soon after you get up. The very act of steeping yourself in these images does seem to assist the chances of bypassing time and seeing any response to one of those items ahead of itself.

5   Dream TV

There is an even more exciting thing you can do with your dream diary experiment. You can try to video the future.

This seems a bizarre idea. However, we have seen that several cases do exist where TV signals have apparently been affected in some way when static-filled screens have depicted scenes from the future. We also saw how George Meek and his 'Spiricom' experiment and the claimed German successes with the related 'Vidicom' programme are supposed to occur because some conscious entity 'modulates' a static interference effect to overlay a meaningful signal on top of that.

Physics implies that future information is coming by way of a consciousness field into our mind and that is able to interact with other energy fields to trigger events at the quantum level. So there is nothing crazy about the suggestion that we train ourselves to effect TV signals.

Here is what to do. During the dream experiment keep a video recorder (VCR) in the bedroom alongside of you. Place it so that your head is as close as comfortably possible to the mechanism. Tune it to a channel that you know will be off the air throughout the night and place a tape into the machine and start to record as you decide to fall asleep. Most machines will automatically cease when they reach the end of the tape. Or you can set the timer to do so for you.

Theoretically the tape should record nothing but static all night long. I would not recommend using anything more than two hours of tape for the experiment as it will become very time-consuming otherwise. When you review the tape in the morning omit the first half hour or so, during which you are unlikely to have been in a deep enough sleep state (i.e. altered state of consciousness) for anything to happen. Fast forward through the rest to see if anything flickers on screen amidst the static. If so, play it!

As I write, a witness in Dorset has written to describe an extraordinary event – seeing a prehistoric scene with dinosaurs on the TV set at 2 am as she and her mother woke from a doze. There were no TV broadcasts on that channel at the time. Imagine, if that had been on a VCR tape instead.

This experiment seems absurd. I have no way of knowing how often (if ever) it may work. But given the TV-based experiences, the Spiricom project, thoughtography-type experiments projecting images onto film, what we know about consciousness and quantum mechanics and the way in which dream states can foresee the future – there is every theoretical chance it could.

Who knows – you may be the first person to video tape tomorrow!

References

Introduction:

Dunne, J.W.: *An Experiment with Time*, Faber & Faber, 1929.

1:

There are many editions of the following science-fiction stories:
Anderson, P.: *The Guardians of Time*, 1955.
Benford, G.: *Timescape*, 1980.
Hoyle, F.: *October the First is Too Late*, 1966.
Watson, I.: *The Embedding*, 1973.
Watson, I.: *The Very Slow Time Machine*, 1982.
Wells, H.G.: *The Time Machine*, 1895.

2:

Davies, P.: *Many Worlds*, Dent, 1980.
Davies, P.: *God and the New Physics*, Penguin, 1983.
Randles, J.: *Beyond Explanation?*, Hale, 1985.
Randles, J.: *Sixth Sense*, Hale, 1988.

3:

Cockell, J.: *Yesterday's Children*, Piatkus, 1993.
Keeton, J. & Petherick, S.: *Powers of the Mind*, Hale, 1987.
Maclaine, S.: *Out on a Limb*, Hamish Hamilton, 1983; and many other publications
Stevenson, I.: *20 cases suggestive of reincarnation*, JASPR, 1966.
Wilson, I.: *Mind out of Time*, Gollancz, 1981.

4:

Green, C.: *Apparitions*, Hamish Hamilton, 1975.
Mackenzie, A.: *The Seen and the Unseen*, Weidenfeld & Nicolson, 1987.

5:

Forman, J.: *The Mask of Time*, MacDonald and Janes, 1978.

You will also find time lapses often featured in the pages of *Fate* and, less often, in *Fortean Times*, both excellent data sources.

6:

Collins, A. & King, B.: *The Aveley Report*, IUN, 1990.
Randles, J.: *The Pennine UFO Mystery*, Grafton, 1983.
Randles, J. & Hough, P.: *Death by Supernatural Causes?*, Grafton, 1988.

Flying Saucer Review also carries reports on occasional teleportation cases.

7:

Evans, H.: *Alternate States of Consciousness*, Aquarian Press, 1989.
Taylor, G. Rattray: *The Natural History of the Mind*, Secker & Warburg, 1979.
Tributsch, H.: *When the Snakes Awake*, MIT Press, 1983.

8:

Grant, J. (pseudonym of Paul Burnett): *Dreamers*, Ashgrove, 1984.
Priestley, J. B.: *Man and Time*, Aldus, 1964.
Zohar, D.: *Through the Time Barrier*, Heinemann, 1982.

9:

Cheetham, E. (ed): *Prophecies of Nostradamus*, Corgi, 1973.
Greenhouse, H.: *Premonitions*, Turnstone, 1973.
Montgomery, R.: *A Gift of Prophecy*, Barker, 1966.
Wallechinsky, Wallace & Wallace: *The Book of Predictions* (part of 'Book of Lists' series), Elm Tree, 1981.

10:

Jung, C. & Pauli, W.: *Synchronicity*, Routledge & Kegan Paul, 1977.
Koestler, A.: *The Roots of Coincidence*, Hutchinson, 1972.

ASSAP often reports on coincidence research and the SPR in a more mathematical and scientific manner.

11:

Capra, F.: *The Tao of Physics*, Wildwood, 1975.
Gribbin, J.: *Timewarps*, Dent, 1979.
Le Shan, L.: *From Newton to ESP*, Aquarian, 1984.
Moore, W. & Berlitz, C.: *The Philadelphia Experiment*, Grosset & Dunlap, 1979.
Schwarz, B.: *UFO Dynamics*, (revised edn), Rainbow Books, 1988.
Zukav, G.: *The Dancing Wu Li Masters*, Hutchinson, 1979.

12:

Randles, J. & Hough, P.: *Looking for the Aliens*, Cassell, 1992.

Randles, J.: *UFOs and How to See Them*, Anaya, 1992.
Rickard, R. & Michel, J.: *Phenomena*, Thames & Hudson, 1977.

Time-travel ideas are sometimes discussed in articles in *International UFO reporter* and *MUFON* Journal.

Fiction References

Books

A Connecticut Yankee in King Arthur's Court, Mark Twain, 1889.
Bid Time Return (filmed as *Somewhere in Time*), Richard Matheson, 1975.
From the Earth to the Moon, Jules Verne, 1865.
Guardians of Time, Poul Anderson, 1955.
Round the Moon, Jules Verne, 1865.
Seeking the Mythical Future, Trevor Hoyle, 1977.
The Embedding, Ian Watson, 1973.
The Lathe of Heaven, Ursula le Guin, 1970.
The Time Machine, H. G. Wells, 1895.
The Very Slow Time Machine, Ian Watson, 1978.
Timescape, Gregory Benford, 1980.

Films and TV

Peggy Sue Got Married, 1986 (film).
Star Trek: 'The City on the Edge of Forever', 1968 (Paramount TV).
Star Trek – The Next Generation: 'A Matter of Time', 1991 (Paramount TV).
The Final Countdown, 1980 (film).
'The Flipside of Dominic Hyde', 1980, and 'Another Flip for Dominic', 1981 (BBC TV).

Sources of
Information

ASSAP (Association for the Scientific Study of Anomalous Phenomena)
20 Paul Street, Frome, Somerset, BA11 1DX

BUFORA (British UFO Research Association)
Suite 1, 2C The Leys, Leyton Rd, Harpenden, Herts, AL5 2TL

Bulletin of Anomalous Experience
2 St Clair Ave West, Suite 607, Toronto, Canada M4V 1L5

Center for UFO Studies (International UFO Reporter)
2457 W Peterson Ave, Chicago, IL 60659, USA

Fate Magazine
Box 1940, 170 Future Way, Marion, OH 43305-1940, USA

Flying Saucer Review
Snodland, Kent, ME6 5HJ

Fortean Times
P.O. Box 2409, London, NW5 2NP

Mutual UFO Network (MUFON Journal)
103 Oldtowne Rd, Seguin, Texas 78155-4099, USA

New Ufologist
37 Heathbank Road, Stockport, Cheshire, SK3 0UP

Psychic News
2 Tavistock Chambers, Bloomsbury Way, London, WC1A 2SE

The Skeptic Magazine
Dept B, P.O. Box 475, Manchester, M60 2TH

Skeptical Enquirer
560 'N' Street South West, Washington DC 20024, USA

SPI Enigmas
41 The Braes, Tullibody, Clackmannanshire, Scotland, FK10 2TT

Strange Magazine
Box 2246, Rockville, MD 20847, USA

Society for Psychical Research (SPR)
49 Marloes Road, London, W8 6LA

UFO Research Australia
Box 229, Prospect, South Australia 5082

UFO Call offers news and information updated weekly at premium rates
(36p per minute off peak as of January 1994):
(Britain only) 0891 121886

Anyone wishing to report any time-related experiences (in confidence if
you request this) or contact the author can do so: c/o 37 Heathbank Rd,
Stockport, Cheshire, SK3 0UP

Index